Education and Science

Education and Science

The Information Processing Age, the Learning Parent and Child in Crisis

Christopher Slaton, Ed.D.

Copyright © 2010 by Christopher Slaton, Ed.D.

Library of Congress Control Number: 2010914189
ISBN: Hardcover   978-1-4535-8405-7
         Softcover   978-1-4535-8404-0
         Ebook   978-1-4535-8406-4

All rights reserved. No part of this book may be reprinted or reproduced or utilized in any form or by any electronic, mechanical, or other means, now known or hereafter invented, including photocopying and recording, or in any information storage or retrieval system, without the permission of the publisher.

The following terms used throughout this book are the proprietary service and trademark of the Progressive Investing Institute of Focused Learning:

Progressive Investing
Feeling System
Systems Feeling
Human Learning Systems
Human Systems Science
Human Systems Research
Building Human Assets
Building Human Asset Meetings
Field of Experience
Community-Based Learning
Family-Centered Learning Plans

This book was printed in the United States of America.

To order additional copies of this book, contact:
Xlibris Corporation
1-888-795-4274
www.Xlibris.com
Orders@Xlibris.com

Without my wife and colleague, I would not have been as able to reach for the capacity and potential to learn how to live with self, other people, and their environments in harmony. Dolores, you have shown me a great deal of care, love, and help.

CONTENTS

Abstract 13
Preface 15

Introduction 19

    Why many poor, minority, high-need, and special-need children cannot learn to receive academics 19
    A Word About the Information Processing Age 22

Chapter 1 A Historical Perspective on Advocacy 27

    Family Literacy 30
    The Family and School Failure 39
    Learning Disabilities 45
    Human Systems Research 46
    Poor, Minority, High-Need, and Special-Need Males 59
    Parent-School Partnership and Accountability 67

Chapter 2 Healing the Brain's Body 70

    The Breath of Life 70
    What Is Language? 72
    What is Cognition? 77
    What is Physical? 81
    What is Mental? 86
    What is Social? 91
    What is Environment? 96
    What is Emotion? 102
    What is Sensory Organization? 108
    What is Green Technology? 112

Chapter 3 Our Thoughts on Family Leadership 117

    Linking Parenting to the Childs Role 117
    Let's Talk about Family Leadership 118
    Now We Can Talk about Family. 119
    Family Language State 124
    Family Cognitive State 127
    Family Physical State 130

Family Mental State 134
Family Social State 137
Family Environmental State 140
Family Emotional State 143
Family and Sensory Organization 147
Family and Green Science 149
Family and Education 152

Chapter 4 A Brief Talk about Our Work 167

Education and Family Improvement 167
The Scientific Context of Parenting Children in Crisis 175
To the Student, Learner, or Participant of Progressive Investing
Practices 179
Safe Places 181
You are in Crisis if you Use Street Drugs 185
About the Facilitators 190
Learning and Training 193

Chapter 5 Parent Education and Resource Coordination Services 201

Introduction to PERCS 201
A Systems-Based School of Thought for Crisis Intervention 202
Crisis Intervention Strategy 203
Community-Based Learning Academies 210
Move through the Crisis 211
Crisis Intervention Conference for Males 213
Sample: One Day Crisis Intervention Conference Press Release 214
Opening Talk 215
The Screening and Assessment Process 219
- Working with Parents and Children 0-12 223
- Working with Males in Crisis CNC Workshop/ Parents
  and Children 10-25 227
- Working with Parents and Youths/Young Adults 13-18 232
Welcome to Our May 22, 2010, PERCS Workshop 238
- Human Learning Goals Influence the Nature of Living,
  Learning, Thinking, and Responding as a Human System 238
- Welcome to Our May 22, 2010, PERCS Workshop 247
- Working with Parents and Children 0-12 247
- Working with Males in Crisis 10-25 252
- Working with Parents and Youths/Young Adults (13-18) 259
- Working with Parent and Young Adults (19-25) 261

Chapter 6 Reading to Your Brain: A New Human Systems Science
to Solve Human Problems..264

- Healing Time..265
- Session I
  - Serve time 267
- Session II
  - Care costs 270
- Session III
  - Feel the world to think 273
- Session IV
  - Trust is a path to respond 276
- Session V
  - Know who you are when you think...279
- Session VI
  - Accept this chance to change the way you act 282
- Session VII
  - Learn to use your brain, to change your mind 285
- Session VIII
  - Stay in the game, to play and to practice................................288
- Session IX
  - Make the brain the home of your ideas to learn how to live ...291
- Session X
  - Is there a choice, or do you just say you have one? 293
- Session XI
  - How long have you been living on hope? 295
- Session XII
  - What do these words mean: Observe, learn, listen,
    help, and lead? 297
- Session XIII
  - Know what the goal is in the aim...299
- Session XIV
  - Look back. Can you feel the help?..301
- Session XV
  - You can do the brainwork, and we can set it up. 303
- Closing Session...305

Chapter 7 A Focused Learning: Human Systems Research Perspective.309

Overview
     Experiencing the Experience of Practice 311
Session 1:
     Coaching 313
     Teaching 314

Mentoring 315
Counseling 316
Team Development 317
Session 2:
Problems of Focused Learning
Human Problems 319
Session 3:
Focused Learning Areas of Human Systems Research
Physics 325
Affects 325
Mentalism 325
Session 4:
Stages of Focused Learning
Stage 1: Focused Learning 326
Stage 2: Aim 326
Stage 3: Progressive Investing 326
Stage 4: Act 327
Stage 5: Sense 327
Stage 6: Feel 327
Stage 7: Focus 328
Stage 8: Discipline 328
Stage 9: Self-Control 328
Step 10: Physical 329
Stage 11: Mental 329
Stage 12: Body 329
Stage 13: Mind 329
Stage 14: Learn 330
Stage 15: Think 330
Stage 16: Social Enrichment 330
Stage 17: Academic Achievement 331
Session 5:
Reflection: Focused Learning 332

Epilogue 337
Systems Feeling Theory 337
Teaching to the Brain 337
The Nature of a Systems Feeling Theory 340
The Learning Parent and Child in Crisis 344
Living Each Day to Become More Informed 348
Parent Education that Works 351

Glossary of Terms 353
References 371
Index 381

Education and Science
The Information-Processing Age
The learning parent and child in crisis

July 15, 2010

Abstract

Our sense is that education has to inform and instruct and that science has to produce new knowledge. In this book, we (Dolores and I) explore the meaning of information as sense data and received data in an effort to explain how human processing leads to information, knowledge, experience, and reflection. This is the Age of Human Systems Science, where the study of how people live, learn, think, and respond describes how we feel things and think through the things we feel. The struggles of a parent and child in crisis are considered, in the sense of how parents need to act as leaders and child developers and in the sense of how children need to act as learners and participants.

When the parent or child in crisis responds to an event it is to become more informed, which begins a rigorous, systemic, and objective word process where valid knowledge may be acquired as they move in relation to a set goal. This makes the act to live each day a practice and the act to learn a process relevant to the affect of education, science, and information processing. Hence, we are using an action learning continuum to capture energy, action, and feelings in the affect through observations of the experiencing of the experiment and the experience of self.

This is what we mean by science-based learning. The systemic collection and analysis of data to learn in relation to historical patterns of inquiry based on the human context of science. According to Eisenhart and Towne (2003), this is the mark of a scientific-based practice in informative research. In this book, we combined participant observations, in-depth interviews, document collection, and evidence-based reporting to explain how parents and children in crisis can learn to move through the experience of hurt. Through the study of learning, behavior, and mental health as stages, we are able to learn from the issues of family decline, school failure, delinquency, and poverty or lack of employability. The underlying dynamics and factors are synthesized through qualitative human systems science.

Preface

Education and science in the information-processing age: Progressive Investing is why the parent and child's act to live and learn together can be seen as a green science. You do the work. To become informed, you have to be able to receive commands that mean to do something. How you do the things you are told to do is called information processing. Because so many things go into how you respond, the outcome or output can be very creative. This is because when your body, brain and senses are set up to work as one system, your response goes through a process to release feelings and/or thoughts. Your body, brain, and senses make up a human learning system.

Find out how Progressive Investing grows the parent and child's capacity and potential to assess their own behavior on the basis of how it affects the way they live, learn, think, and respond. The practice of Progressive Investing takes place in four environments that offer feedback on the parent and child in crisis act through a continuum. We gather objective data on what works, where it works, and why it works. Hence, the more the parent and child work in pursuit of their goals along the continuum, the more they are able to practice learning how to become more informed, knowledgeable, experienced, and reflective.

Progressive Investing models encourage and support parents who strive to reach their potential to meet the needs of their children through a vision, signs of care, and aim.

Progressive Investing Model 1

Progressive Investing ⟵———◯———⟶ **Green Technology**

Practice

To go further, the acts to live, learn, think, and respond in the practice of Progressive Investing lead the parent and child to gather a more in-depth sense of what we mean by green technology. The change process in each step forms the vision, signs of care, and aims we have identified for the information-processing age to improve the parent's ability to lead and develop children, and the child's success as a learner and as a participant. Focused on the four social places we all share life experiences—but apply this focus to improve the things they do in home, school, neighborhood, and workplace situations—to learn how to live each day to become more aware.

We wrote this book to help people in crisis learn how to act through it. Every human being has to learn how to use the body and brain to act through times of pain or pleasure. How do you feel things, as feelings move through your body, brain, and senses? You are about to feel how we work with a parent and child in crisis to learn ways to help. In this sense, we are here to help parents protect their child's right to learn how to grow as a green science. This is the human side of learning how to use space, energy, and time to help improve the way self and other people use the environment to live, learn, think, and respond.

When you think, how you feel pulls from the act itself. This is why when a parent and child that have been hurt by major life events—which affect how they live, learn, think, and respond in home, school, neighborhood, and workplace environments—take action to learn how to move through it, how they sense, feel, and focus their brain and body is in part a green science. This means they are Progressive Investing in the act to learn how to reflect—look back—to regain the human ability to feel through the affect of an experience, to help.

Chapter 1 provides a historical perspective on advocacy, which explains why many poor, minority, high-need, and special-need children cannot learn to receive academics. In other words, when a child does not appear to know how to receive information with intent to organize a sense of it, we explain what this means. According to these circumstances, "Kids do not fail, schools do." Through this pilot human systems research (HSR), academic failure was linked to anger, perceived lack of ability was linked to fear, perceived unfairness was linked to anxiety, poor school experiences was linked to lack of academic success.

In chapter 2, we talk about the breath of life in relation to human contact and a baby's initial acts to sense, feel, and focus brain, body, and sense events. Since this is the stage where many children are hurt by the way they are received, this chapter is called healing the brain's body. How a baby feels contact sets up their sensory system backward to inform them as acts to think forward. Thus begins your baby's search for understanding contact, interaction, and signs of cooperation.

Chapter 3 talks through our thoughts on family leadership, because as grandparents, we continue to be concerned with how a child develops. We explain how parents learn to work with us. Parents use the role of parenting to perform in task that lead and develop a child as a learner and participant in their family system. We talk about the needs of children who live in broken families. This may be a child who lives in a fatherless home. This may be a child who needs to learn how to make choices. This may even be a child who has to learn how to work through the way he or she feels to be upset, to learn how to feel for self and others.

In chapter 4 the reader is asked to review a brief sketch about our work. Such as how we help parents read and interpret their feelings as they move their children through puberty. We can observe through the use of talk and action how parenting unfolds as a humanistic and social practice of becoming reflective in the study of a child's response. Many parents and children in crisis have to learn how to move through day-to-day events that are threats to his or her physical or mental well-being.

Chapter 5 explains the Parent Education and Resource Coordination Services we offer to improve learning and support services. For instance, using Building Human Asset Meetings with parents, children, teachers, school principals, and political leaders is an objective way to gather data to learn and understand the problem of family decline, school failure, delinquency, and poverty or lack of employability. This section bridges all the systemic components into a series of research-based interventions to improve learning and support services to all participants.

Chapter 6 explores the use of words to explain the purpose of reading to your brain. The brain is wired to sense and receive data to improve how you read and interpret words and their meanings. These words are meant to be played with, tested, experienced, and researched to guide your journey to a healing perspective. We use research-based approaches to learn the power of words you need to feel to focus the experience of talking to your brain. Each session provides a new approach to learning how to recover from hurt through sets of words that talk to your state of mind.

In chapter 7 we discuss what focused learning means. In focused learning, when the brain and body works in concert, he or she is called a human system to account for their mastery over the brain's body. Get to the heart of why the sense, feel, and focus cycle is set up to help the learner learn how to manage the flow of energy. For instance, the learner learns that energy flows through the body, but when the brain is the lead, the energy becomes more controllable. In focused learning, teaching is the experience of learning how to use contact to study the act to sense, feel, and focus the aim to care.

Finally, we offer a general systems feeling theory that draws from a synthesis of these new sciences, as a new school of thought. Space is never empty, time is never stopped, and energy is never held back by the same rate of care. This means, the qualities that make up the spirit of the earth are global. Life is such that we can find parents and children all over the world out of touch with each other, as though unable to make sense of each other's human act. This is the process we describe through human systems research as forming practical ways to study self action that relates to the acts of other people and their actions in the environments they all share as a man-made system; homes, schools, neighborhoods, and workplaces.

Christopher K. Slaton, EdD
Human Learning Consultant

Introduction

Take the next five seconds or so to feel your space. Let this time be a part of your fate. Now is the time to think through these words. Do not resist how your brain, body, and senses receive this work, all at the same time.

Why many poor, minority, high-need, and special-need children cannot learn to receive academics.

On July 3, 2006, we submitted our initial report on why poor, minority, high-need, and special-need child cannot learn in relation to drug-related disorders. We requested Building Human Asset Meetings with the State Superintendent of Public Instruction to discuss ways to improve the lives of disadvantaged parents and children who rely on public schools for family development and support. From now on, we want to amend that report to include "why many poor, minority, high-need and special-need children cannot learn *to receive academics.*"

The process of learning how to learn starts in the home and involves the experience of contact that the child is allowed to explore, to find out what it means to be human. Being human is both an action and a process that has to be learned. Every move a child makes is an act in relation to contact and a process in relation to interaction with an event. The event of learning from the act to make contact and the process to interact leads the child to feel and respond to an affect. The result of which moves the child through stages of contact to states of interaction as events of perceptual, cognitive, and physical learning.

When a child reaches school unprepared for contact, they naturally will move in relation to an aim to interact with other children, but not to learn from the way they sound, feel, or look. Children that are unprepared for school go to school, to make contact and to interact, but their aim is to play. On the other hand, children who reach school ready for contact will have a look of focus about them that is more aimed toward the adults and will learn to interact from the way they sound, feel, and look. These are signs of awareness for the role of school through their human resources.

One child goes to school to learn how to use their body, and the other child goes to school to learn how to use their brain. The way humans learn how to learn leads their action (Argysis and Schon 1974). A child is not set up to learn academics when their body and brain's relationship to each other is hidden. For instance, when a child is not set up to learn academics, they have a pattern of behavior that signals an alarm that mean "I am not ready for contact because I do not know how to interact." When a child is set up to learn academics, they have a pattern of performance that signals "My senses are armed to help my brain learn to control my body."

Our concern is for the child that is behaving, because these are signs of a child in crisis. They are more resistant to contact and interaction. They display a lack of internal control because their brain, body, and senses are not set up for them to experience academic events. They have a harder time sitting still and controlling their body movement and mental states. All these things are known through their first day of school. The nature of their brain, body, and senses are hidden resources in that initial perceptual, cognitive, and physical experience of school.

In other words, when a child does not appear to know how to receive information with intent to organize a sense of it, it means they are less prepared to communicate. Goleman (1995) studied emotional intelligence within the meaning of our interpretations of how a child needs to feel through experience. It is at odds with the schools need to shape the process to program the event to learn, which affects how feelings are valued. Durkheim (1963) wrote long ago, that nothing in society is real; accept (I will paraphrase) the child whom is living to learn how to live in it.

When a child is not able to pick up the noises, sounds, signs, and symbols that make up the academic curriculum they appear not to be ready to learn. Our perceptions of cognition flow from studies of Piaget's work and Maslow's thoughts on how children learn through the experience of action, which leads them to process the act of acting through the behavior. This means that the child who comes to school already in a fixed mental state has less of a chance to learn how to move through it because of emotional distress.

Kuhn (1962) wrote that the American learning system is designed to create confusion between people, based upon the schools they attend, to learn how to practice learning from other people (Slaton 2002). A child in crisis will appear to be unable to profit from instruction when their behavior becomes less stable as a result of emotional stress. This may be called a learning disability if the child displays a lack of ability to receive, organize, and communicate through the use of academics. Academics begin with being able to sense, feel, and focus brain and body events. The human senses are referred to as the child's paths to knowing what is in the world (Mueller 1965; Slaton 2002).

Not being able to read at or above grade level is a crisis. Not being able to speak at or above grade level is a crisis. Not being able to listen, not being able to write, and not being able to count reflects a child that is not set up to receive sense data. As we reported to Congress in 2006, these children are in an emotional crisis because their brains are being damaged. We have met with representatives of the chair of the Education Committee, U.S. congressman George Miller to submit our report. We have met with the chair of the Health and Human Services Committee, State Assembly Member Jim Beall Jr. to submit several of our reports.

The perceptual and cognitive aspects of being human—the affective aspects—the action and process aspects of learning how to move through contact and interaction are all at risk. Segal and Yahraes (1978) applied the label to children in crisis, as is described above, to have a "psychiatric disability." As he or she moves on into their school years, they grow more frustrated, disappointed, and hostile toward themselves and others. They tend to drop out of school and end up in prison, because they lacked academic success in school.

Hence, we have called for early screening and assessment of learning problems. Dewey and Bentley (1949) are another set of researchers and scientists who wrote long ago that word designation is an act of science. The word *psychiatric disability* was meant to cause a professional dialogue through the applied use of these words in relation to learning how to work with the child's brain in mind. Wells (1999) would call this kind work, an effort to produce thought through the use of talk, to talk to each other's brain.

A WORD ABOUT *THE INFORMATION-PROCESSING AGE*

The title of this book is *Education and Science: The Information-Processing Age*. The purpose of education is to inform the brain, body, and senses on ways to act with care. The purpose of science is to arrange a body of knowledge that is coordinated and systematized (Murphy and McMahan 2000) to sense, feel, and focus a person's labor through the use of speech and written words. The purpose of information is to set up the event to receive contact (energy, action, or behavior) to learn how to live through the experience of feelings for a high sense of self, other people, and the environment. The purpose of processing is to experience the interaction of these feelings through patterns of learning, thinking, and responding to contact. The purpose of this age is to experience ways of knowing how to coordinate acts to live, learn, think, and respond in complex home, school, neighborhood, and workplace networks all at the same time.

How information processing takes place between brain, body, and sense events is not regularly viewed as a field of modalities. Just as the human system is set up to experience language, cognition, physical, mental, social, environmental, emotion, and sensory as green sciences—to explore the domains of human systems research—we explain the ways in which the brain, body, and senses process information in the experience of knowledge. We see the human system as an information processor that lives through the use of talk, mechanisms, matter, mind, contact, man-made places, feelings, nerve networks, and ecology. Hence, language, cognition, and human physics are described as the necessary order in which we set the stage to become mental, social, environmental, emotional, sensory, and green technologies. For example, teaching a person how to talk sets them up to structure and improve the way memory functions feel, and may focus a physical sense of self-awareness in the process of becoming more mental.

We believe this method of application sets up treatment structures that function to help people that have been hurt by major life events recover through their interdependent relationships. They become more responsive to tension, stress, and pressure as a human system. Human Systems Research is used to study certain process variables and changes not openly used to learn how the body, brain, and senses grow, mature, and develop with a capacity to recover from being hurt. This is how the use of speech, recall, and bodily responses help a person that is or has grown up hurt to reduce gaps between a lack of feelings for self and others and a high sense of feel for self and others. To become more informed means to become educated. This in turn means to improve the labor of a person through the use of information. The science of learning how the person processes information lets us look at interdependent structures and functions that are examples of acts to live, learn, think, and respond across a continuum.

The staging of information processing scales throughout this book openly seeks to inform, with an emphasis on the affect in experience, on acts to move in relation to goals. The information-processing age will be determined by the interaction between the act to inform and the practice to perform. A successful process variable is one that helps the person to grow through the experience of contact, to learn from the experience, ways to improve their labor. The words we use are intended to apply, perceptively, to the human system, any state of becoming, or any condition of acting, regardless of when, where, how, who, what, and why the hurt blocks. The science of learning how to live through the experience of contact has been discussed.

The main purpose of this book is to introduce the kinds of things studied by a human, cognitive, and behavioral school of thought and to a larger degree, Progressive Investing—to live each day to learn how to become more informed—through the act to feel the pain of being hurt—is used to set up the events and goals in each day. This provides the context for understanding real life problems in their present forms. This is why we will focus on ways to improve being human through states of awareness in the situation of becoming an actor in home, school, neighborhood, and workplace networks that influence pain and hurt.

We are here to help. All complex entries into life may produce mechanical blocks in the growth, maturity, and developmental cycles of becoming a human system. Understanding how words may touch an infant's brain helps us to understand how initial cognitive structures can build up a resistance to contact when the infant's physical acts to live are not realized as feelings for interaction. Understanding these three basic stages to becoming comfortable with self and other people in a new environment relate them to the function of mental capacities. Furthermore, these feelings to become mental lead into the social path of interaction in which signs of environmental awareness sprout with emotion to be shaped through sensory organization of send and release cycles that form this new green technologies special sense of reality. Though your understanding of information processing cycles may be different, on this point we may agree, it is the goal of life to understand these things.

This book is about the basics of information processing, to learn how to live with self, other people, and their environments. It suggests that growing up hurt is a major life event that deals with three basic process cycles, which we call sensing, feeling, and focusing brain, body, and sense activity; receiving, processing, and responding to contact, interaction, and cooperation; and living, learning, thinking, and responding in home, school, neighborhood, and workplace networks. Communication deals with the acts to transfer information between self, other people, and their environments (Kuhn 1975). In all cases, feelings are exchanged through physical and mental paths of awareness unless blocked in the transfer of the experience. For instance, a sender's body may be physical, but the noises or sounds they use to communicate may be intangible and yet mental. Because of these aspects of feeling, the contact through the senses, how it feels to interact can be blocked by the effect, which emerges from the experience.

Information-processing cycles are set up by the way the brain receives the sign or symbol, or noise or sound as a building block for future experiences. This is why poor experiences with contact have a certain effect on the physical act to become mental, when there is a resistance to feel for the event of social interaction. In fact, you could conclude that emotion is blocking the aim to receive what has been sensed through the act to live. The transaction deals with the transfer of feelings across a continuum of sending and receiving messages to learn how to take the next step. Anger, fear, and anxiety—this arrangement speaks to the effect on emotions that may block the process cycle. Negative cognitive structures reduce or build a resistance to the experience of contact for fear of being affected by the interaction. Hence, acts to become physical and mental are not realized in the effort to organize a sense of these feelings, in that order.

If you are here to learn or to help, then make the parenting of your child a practice in the way you experience learning at home. This sets up the picture. How you learn with your child is a strategic practice to plan and assess the way he or she will move from home to school as they live through the K-12 system. The truth about how they learn, behave, and conduct themselves will come out through the way they use the senses to receive data. As a parent, you really do not know what is being taught to your child in a classroom unless you are there. But you can set your child's brain and body up for the experience of school and the possibility that there will be challenges along the way. We offer a helpful approach to learning how to use your home as the base to learn how to deliver advice, advocate, mentor, and lead self, and other people to seek a deeper sense of understanding for your child as a learner and participant.

We are asking you to learn from us and with us, to set the process up to lead and develop your child as an asset. Using evidence-based reports, we want to share a historical perspective on our work with children and parents in crisis as their advocates. This will directly link our work to the big picture: the way we move a child that has been hurt by major life events in their home, school, neighborhood, and workplace network through the K-12 system improves through learning and parenting with strategic action. In the Information-Processing Age, you, and your child need to be able to sense and receive data through process cycles that reduce any signs of emotion that may affect the flow of data to and from the brain.

Chapter 1

A Historical Perspective on Advocacy

The models of this book began to consume my thoughts over twenty-nine years ago. At that time, in 1981, I realized I could neither read nor write well enough to represent myself as an adult. I was in the worst of spirits when my older brother introduced me to a new friend—his sister-in law, my wife, Dolores. As a new friend, my current journey and sense of now began on that day. I recall how my desire to learn how to act more intelligently was groomed from this initial meeting.

What Dolores brought into my life—right from the start, her reading and writing skills were being put to the tests. What kind of person was I to move someone else to do my work for me? Well, Mrs. Johnson, my college English grammar professor, caught on. If you fail to prepare to work in her class, then you had better be prepared to fail. In other words, if I could not do the work in her presence, then I was not prepared for the real act of doing the work. She failed me, with a crystal clear F, but told me "You can do this work."

In the twenty-nine years since Dolores became my friend, research aimed at answering the question of what it means or feels like to grow up hurt by major life events in your home, school, neighborhood, and workplace network has moved me. It has long seemed to me that no subject warranted the attention of a child and family advocate more than the knowledge of how to help. I am in part, describing a state of mind as being in crisis, because no one had ever talked to my brain the way Dolores and Mrs. Johnson did back then. It was as if they were messengers seeking to recover me from my own hurt and backed-up feelings.

For many years henceforth, I have written research-based reports dealing with one aspect of growing up hurt and the other aspect of being a feeling person, hoping all the while to one day be able to make sense of my personal school experiences, which have shape the focus of my work. Having grown up in a broken home, school, neighborhood, and workplace network, my aim is to produce products and services to improve learning and support services to children and families as they move through similar states of crisis.

Every parent and child that has come to me for help, I have worked with as though they were my mother and me. Grounded on research-based principles for all of us who refused to hear their mother's voice, cut school in search of ways to stop the hurt, used their neighborhood as an escape from authority, or rejected the helper's help. In between these lines grew a child, youth, young adult, and adult who was learning how to think, but not how to feel things. In advocating for parents and children, I feel as if I am advocating for the child's right to learn how to feel things and to learn how to think.

The opportunity to work for parents did not emerge in my practice until 2004, when a child in crisis was brought to my office by a parent wanting to know what her child was feeling. Dolores, now my research assistant, reported from her research notes that the parent had been threatened by school officials to call the police whenever the parent was on campus. She was considered an out-of-control parent. "The concept of being in crisis may have many meanings. These meanings change depending on an individual's past experience and his/her understanding of stresses in everyday living" (Slaton 2002, iii; Harbert 2000, 385).

The task of producing a child—and family-centered learning model, enormous in its scope and complexity, could not have been completed without the help of parents and children wanting to learn how to receive help. A Progressive Investment Report (PIR) was created using our year round meetings and talks with children, youths, and young adults. In the interests of family research, parents and children came forward to submit their official request for help on May 26, 2005. Part three of the report read, "Human, cognitive, behavioral problem solving for families in crisis as a special service" (Slaton 2005, 9).

Meetings were conducted with parents, children, teachers, school principals, student study teams, and so on to investigate the need for the services requested by parents in crisis over their child's school failure. How does a child feel when they are failing in school? What sways the choice to act or how to act for the parent? Is it the family—the single parent crisis—or the rise in drug use? We talk to people to learn firsthand the quality of family life, the stress and pressure of school interaction, or is there an elephant in the neighborhood? As human systems researchers we strongly believe there is a pattern to follow.

What is it that causes a parent and child to disconnect and isolate, withdraw or stop being friendly toward each other? More importantly, why does each of these cases lead us back to the school? We wrote over eighteen letters to the school district in 2005. We received ten responses that did not passably respond to the parent's requests for improved learning and support services. School officials acted as though they did not understand how a family relationship could be at stake, over poor school experiences. For instance, one parent was offered classes by a consultant for the district—the agenda topic read, "Life in the Family Zoo."

Of course the parent felt insulted and wanted to know why the school would be so insensitive. Another parent was struck by the school's response to her child's behavior. For instance over a three-year period, the child would be removed from the classroom and sent to the office, where he would sit and remain for the school day. A husband and wife grew at odds with each other over what to do about their child being constantly bullied at school. In each situation, the parents' stories became more credible as we continued to meet with school officials at their school sites and the district office.

We labeled these threats in accordance with the education code, as psychological and oppressive acts. On June 4, 2004, an official request for an investigation was made to the district and the county office of education. Another example of an oppressive act: One parent had submitted a written request for an individual education program (IEP) meeting in 2002, when her child was in the second grade. The parent did not receive a response until 2005. Ms. Zeeman, director of elementary education, wrote, "The purpose of my correspondence is to provide a few specifics related to our meeting on November 10, 2004. We are committed to making critical changes in our practices and in the way we relate to students and their families. These changes reflect our acknowledgement of a serious need to ensure that our students in all racial, language, and social groups feel safe, welcome, and successful at school." In our view, the work of education has to be about the needs of the students and the families they serve.

Family Literacy

On June 3, 2004, we filed our request for an investigation. The purpose of our complaint was to request an investigation of the district's human dignity and civic discourse policy. The parents in this case were being discriminated against based on a lack of learning and support services and social profiling, and this led the school to ignore request for information to consider how their children were being treated with education. The parents' local schools did not comply with specific sections of the education code and, by failing to respond, used psychological oppressive acts to reduce the effectiveness of parenting, advocacy, and education on behalf of the children and parents in attendance.

The school historian Lawrence Crimin (1990) reported on the problems of achievement in relation to changes in family. "For one thing the family was the institution in which the child had the earliest education, their earliest experience in the learning of languages the nurturance of relationships, the internalization of values and the assignment of meaning to the world" (p. 53). We filed a complaint with the school principal, Mr. Bonta, on December 11, 2004, concerning the "continuation of school interaction problems" (Slaton 2004, 55). Children first develop the structure for personal, academic, social, and occupation learning styles in the family. We have been studying public schools for more than twenty years, and from this point on, we were shocked.

These parents were begging school officials to help them become more informed, but because their request included changes in assessment, learning, and teacher practices, they would continue to be ignored. From this perspective, learning and behavior responses of children in crisis, for this group, reflected those of a culture of children who methodically reject school and teachers (Ogbu 1978). The parents described these children as falling through the cracks, because their families had been destroyed by the crack cocaine epidemic, three strikes sentencing laws, and unemployment. The reasons for these continued declines. We reported rises in teen pregnancy, poverty, mental retardation, and maladjustment disorders and emotional disturbances. The children ranged between infancy and age twenty-six.

According to these circumstances, "Kids do not fail, schools do." Through this pilot human systems research (HSR), academic failure was linked to anger, perceived lack of ability was linked to fear, perceived unfairness was linked to anxiety, poor school experiences were linked to lack of academic success (Slaton 2004, 56). In children, especially youths, anger is an impulse to act out, fear is an impulse to feel for safety, and anxiety is a trigger that affects the senses, emotions, mental processing, physical behavior, and social displays (Goleman 1995). This was a middle school youth. On October 23, 2004, we assessed four unsolved cases. Children who do not experience academic or behavior success in their school years often become truant or drop out, and even if they finished school, they are inadequately prepared for life tasks and are assured underemployment, long periods of unemployment, and dependency on public welfare or other subsidies (Caple 1990; Slaton 2004).

Many underachieving students feel trapped in schools where teachers and school principals just don't understand them (Kuykendall 1992; Slaton 2006, 60). According to Kuykendal, "The key ingredient for improving the academic self-image of all youth is accomplishment" (p. 23). School achievement helps a family in crisis, understand school affairs. Building Human Asset Meetings (BHAM) were held, as the process to link community-based action research to the people who share or are responsible for providing leadership to the parents and children's home, school, neighborhood, and workplace networks. The purpose of a BHAM is to improve learning and support services to poor, minority, high-need, and special-need children and parents. The information that is gathered in a BHAM is fused with practical advice to inform and instruct decision makers.

In BHAMs with several school officials, it was clearly stated this is not a child's behavior that makes him or her difficult to function in a classroom. In our laboratory school, parents and children participate in HSR-based practices to learn how to improve their home, school, neighborhood, and workplace relations. This was a child without an agreed-upon 504 plan, one whose parent has requested additional tests, and one that is in crisis, because of being held back a grade (Slaton 2006, 61). In subsequent meetings with school officials, we asked them to explain why this child was not achieving at the specified levels of performance. M. Guzatis, the acting school counselor, failed to bring the student study team back together in a timely manner and, hence, failed to provide the child and parent with the requested support services.

In another critically unsolved case, we met with program specialist Ms. Alhmann concerning a June 18, 2006, IEP meeting. The parent continued to express a difference of opinion with the school psychologist. The parent reported her child did not learn to walk until after eighteen months of life. The parent conveyed that her child was unable to climb stairs at thirteen months of life and had difficulties using his hands to grip certain objects. The parent pointed out these reports must be accepted, because they relate back to her child's gross motor developmental milestones. The parent reported in the area of expressive language milestones that she had to learn how to help her child learn to use his mouth to talk. The parent's report specifically reflected on a child with communication problems in need of training to use and understand language, gestures, and other people's facial expressions. Our school site visits revealed the child was isolated by peer teasing and had a difficult time adjusting to people, objects, and events (e.g., talking calmly, moving in and out of contact, and adapting to changes in routine or familiar surroundings).

The parent advised the IEP team members of these difficulties, hoping to learn from them how to help her child through the use of education. In this case, the mother was asking school officials to inform her concerning the things she could do. She explained how going back to school for herself directly influenced her ability to provide better care to her son. What was she to do? Did she really need to admit that she had been a drug user while her child was in utero? The parent displayed problems paying attention, signs of being hyperactive, as well as a difficult time listening and preparing thoughts. The parent talked about personal interpretations of her outward behaviors in relation to her son. The IEP team of school-based officials did not agree on the classification of the child's problem as a conduct disorder.

The parent and the advocate verbally disagreed with the rest of the team. This stirred more emotion from school officials. A motion was sought to rearrange the child's pre-existing 504 plan, even though we continued to disagree, based on the premise that family history was grounds for an expanded assessment to address how the child's academic achievement gap would be reduced. School officials were not basing their position on grades and tests scores, since the youth had been failing classes over more than a two-year period. Three years of failing grades are a signal that the child lacked an ability to receive, organize, and express information. This is a learning disability (Pierangelo 2003); an academic disorder (DSM-IV 2002); signs that the sense and receive process cycle is hurt. Add to this point the fact the youth was constantly being teased and bullied over this span. The behavior pattern clearly reveals a form of school phobia, disconnection, and withdrawal from school events.

The levels of frequency in the child's classroom conflicts and interaction problems point toward serious human relations issues, which has had an effect on the parent's child, and his ability to learn and feel safe in school. Hence, this issue, of how a child grows up hurt by major life events in their home, school, neighborhood, and workplace network has been framed. So far, we have covered five cases. In four of these cases, there was a single mother working with a fatherless child, and substance abuse was involved at one level or another. We do not pick our cases. Parents or children find out about us, and one or the other will call us for help. In other words, family relationships and parent education for these parents had to be more than academic counseling to meet the daily challenges they were facing. Parents realized what it felt like to be in crisis.

Family literacy was what the parents wanted to attain through their local schools. Their responses were not forthcoming in this regard. School officials were not aware of, or lacked knowledge of, how to talk to the parents and children's brains. The parents and children were in this struggle together, because they were learning why they tend to do the things they do. Plain and simple, we used words of care to talk to their brains, to learn how to help, and to meet their needs. To reduce their child's signs of mental confusion concerning home-to-school learning, we held weekly workshops. BHAMs were held—one for parents only, one for children and youths only, and one for the families. The parents asked the district to look at the problems as one that involved school safety, in that, the human relations problems did stimulate emotional responses.

On June 1, 2006, a complaint was filed. We reported that the school had violated the parent and eligible (IEP) student's due process rights through the use of trickery and deceit to move the child through the year without providing formal (adequate learning and support services) responses to the parent's request for amendments. For instance, the parent objected to the school psychologist Mary Yaryan's attempt to present a Pshycho Educational Evaluation that was conducted by the child's elementary school dated May 9, 2005. This action violates Public Law 93-380 of the Family Education Rights and Privacy Act. "Parent and eligible student who believe that information in the education records is inaccurate or misleading may request that the records be amended" (Slaton 2006, 79).

The parent informed each IEP team that Anthony has unique family, medical, and educational problems, because he was subjected to street drugs while in utero. As an infant, his mother who used crack while pregnant exposed Anthony to the effect of cocaine in utero. Because of this, the parent has requested a more specific classification to address both aspects of his academic and behavior deficits. School personnel representing the middle school were advised of this request before the start of the current school year in a meeting with the vice principal and again on October 19, 2005.

A meeting was held on May 11, 2005: Triennial IEP meeting. Severe discrepancy exists between ability and achievement in basic reading, reading comprehension, written expression, math calculation, auditory processing, and sensory-motor skills. Based upon a review of the information referenced above, the members of the IEP team have determined that additional assessments were needed in the following areas: health, academic development, intellectual development, and social or emotional development. Hence, the school psychologist erred when she did not provide new information concerning what the IEP team determined on May 11, 2005. Hence, her determination that Anthony did not qualify for critical care as a developmentally delayed student was the red flag.

We met with Sue Stickel, deputy superintendent for the California Department of Education on April 4, 2006. We explained, "We continue to be very troubled with the tactics that are being used to neglect or ignore our request for an investigation." A copy of our 2005 report had been mailed certified to Ms. Johnson, chair of the state board of education. In support of our request for an investigation, we have submitted evidence-based request for an investigation in accordance with section 14000 of the education code. "Local school districts should be so organized that they can facilitate the provision of full educational opportunities for all who attend the public schools. The system of public school should prohibit the introduction of undesirable organization and educational practices, and should discourage any of those practices now in effect" (p. 83).

State Superintendent Jack O'Connell wanted the matter to go away. Our BHAM proposals, we explained (May 15, 2006), are remedies to problems that affect children and parents throughout the state of California. Our proposals are especially for students that have been hurt and injured. The superintendent was provided with physical and written evidence that HSR had been carried out with rigor, that the procedures and processes of report writing had reduced the possibility that the investigation was superficial, biased, or insubstantial. For instance, a PIR is derived from interpretive research-based practices and is presented as HSR. Like other forms of interpretive research-based practices, HSR seeks to reveal how humans live and learn through home, school, neighborhood, and workplace experiences.

Hence, we provided cases of children and parents in crisis to enable the state superintendent to sense and feel how and why issues and events around home, school, neighborhood, and workplace relations have an effect on their daily lives. Crisis: There are 2.5 million grandparents raising their grandchildren; 960,000 of these children have no parent in the home (CDF 2007). This is a war against child poverty, which cannot be won without a focus on ways to teach fathers how to live in a home, with their children in mind. Hence, poverty is the leading connector behind broken families, broken courts, and the "school to prison pipeline" crisis (p.25). Our 2006 PIR identifies thirteen counties in California that have been hit the hardest by the drug epidemic (p. 11). Children with drug related disorders are more likely to be carded with conditions associated with behavior or conduct disorders, and with no reference to organic or mental causes, regarding how and why, the things they do have a negative effect on cognition, emotion, and development.

At the most basic level, a crisis is defined as a temporary state of upset and disorganization, characterized chiefly by an individual's inability to cope with a particular situation using customary methods of problem solving, and by the potential for a radically positive or negative outcome (Reineke 2002). The cognitive-behavioral approach takes a more expansive and inclusive approach by focusing in greater depth on family interaction patterns (mother, father, son, daughter; brother, sister) and remaining consistent with elements derived from a systems perspective (Datillio 2002).

In 2002, mothers under the age of twenty gave birth to almost 850,000 children. Many of these young mothers were hurt or injured by the conditions of their neighborhoods while growing up and are less equipped to live and learn how to use human—and land-based systems. We know how teen parenthood increases the risk of generational declines in home, school, and neighborhood living and learning situations. For example, every year, more than 600,000 youth are detained in secure facilities. The exposure of young people to incarceration contaminates their chances to live through and learn through these challenges. In 2005, California admitted 9,046 juveniles into the juvenile hall system. Of that number, 3,400 received mental health services, and 1,219 received psychotropic medications. According to federal statistics, on any given day, there are approximately 27,000 youth in juvenile detention centers. This is an increase of over 100% since 1985. Many of these youth are parents (PIR 2006, 18).

Social workers have found drug-abusing mothers to be extremely difficult to involve in meaningful planning for their children's future. Parental contact is erratic and seldom sustained, which is a measure of the mother's fitness or levels of disability in relation to parenthood. The mental and physical breakdown of the mother is simply seen as not paying attention to the baby's cries for help. The behavior of a parent that cannot function is clear. What is not so clear, however, is how the home, school, neighborhood, and workplace network is often the systematic link to the parent's downturn. These environments relate to the child and parents social, academic, legal, and employment status, from which emotional conflicts emerge.

These multiple problems reflect how insufficient the support system actually is for low-income children when the child is placed at risk, because of neglect or abuse due to the physical or mental illness of parents, a broken home, or unmarried parents since there is usually insufficient help to be obtained from other relatives and/or adults. Hence, the child, because of unmet needs that grow to be more and more complex, rejects the other people in the network (parents, teachers, principals, social workers, and so on).

In 2003, 74% of adult males arrested tested positive for drugs or alcohol. Drug abuse affects 40% of newborns before their life begins. Conservative estimates suggest that at least 11% of all newborns in the United States today were exposed in the womb to one or more illegal street drugs. Children exposed to drugs while in the prenatal stage live in the same neighborhoods and go to the same schools as children who were not exposed. However, their home, school, neighborhood, and workplace environments differ in this important respect: many of the mothers of drug-exposed children relapse at least once after giving birth to a child, and 60% continue to abuse drugs.

For instance, Jenkins and Sauber (1966) interviewed 425 families in New York City as soon as possible after their children had entered public foster care for the first time. When parents were asked to tell what happened that brought about their child's foster care, they described a variety of crisis (Costin, Bell, and Downs 1990). Several of these important studies began in the 1960s or early 1970s, with results reported in relation to socioeconomic status for children at risk of school problems in the late 1970s and early 1980s (Erickson and Riemer, 1999). Studies of children with Down syndrome are more common and easier to link to the problem of substance related disorders because the syndrome is easier to identify (p. 13).

Unlike alcohol, researchers have not been able to associate any pattern of birth defects with cocaine, but note that drug-exposed babies are more likely than others to have family, medical, educational, and social difficulties. Basic psychological processes such as cognitive functions and associations are affected by problems living (Szasz 1974). Drug-exposed babies appear normal in the sense of intellect, but show significant problems living with other people because of poor internal control. They appear to be more impulsive, easier to distract, and show rising signs of aggression.

Mental illness of this type and character has many different etiologies and may be seen as a final common pathway of various pathological processes that affect the functioning of the central nervous system. Mental disorders, then, are physical and biological processes that cause behavioral factors that are in part due to a person's general medical condition (DSM-IV 2002). In 2003, 74% of adult males arrested tested positive for drugs or alcohol. Three-fourths of incarcerated youths have been found to suffer from mental health disorders, and one in five from a severe disorder. Boys are five times more likely to be found mentally ill than girls are, even though in 2003, almost fifteen thousand girls were incarcerated (CDF, 2007).

Poor, minority, and high-need children: If the learning and support services were in place, the superintendent of Public Instruction (CDE) was asked to inform us (PIR 2006, iv). Children that have been hurt by major life events consistently fail in at least two critical areas of learning (1) in the acceptance of contact and (2) in the act to interact. They display a lack of internal control because their brains have been injured (p. 5).

In 2000, approximately 550,000 children between the ages of sixteen and eighteen were in need of foster care services. Of this number, black children were disproportionately represented. They were reported as making up more than 40% of the foster care population, even though they represented less than 20% of the nation's child population. One in two children will live in a home being parented by a single parent. One in three children are born to unmarried parents or an abandoned parent. One in three children are behind by at least one year or more in school. One in four children live with only one parent and that parent's level of education will determine how successful they are at moving through the public school system (Children's Defense Fund 2003).

In September of 2007, we reported to the Deputy Superintendent (CDE), "The large numbers of single parents among African American families, the high rates of substance abuse and their disproportionate contact with the criminal justice system places mandatory consequences aimed at them in relation to incidents of reported abuse and neglect. For instance, 65% of African American children were removed from their homes because of parental substance abuse, compared to 58% of white children [Government Accounting Office (GOA), 2007, African Americans in Foster Care]."

Group homes are operated like detention centers, in that they are highly restrictive environments. Approximately 43% of out of home care is provided in connection with the department of probation. Children that end up in foster care are more likely to be sheltered by a relative 32% of the time or a certified foster care agency 25% of the time. Approximately 57% of the costs to care for all the children in-group homes were covered under child welfare or Aid to Families with Dependent Children funds. In 2000, child welfare agencies supervised more than 90% of all children in foster care, which is separate from the 57% in-group home placements.

Hispanics make up 41% of the general population, and 58% of those in poverty. In contrast, blacks only make up 7% of the general population and 8% of those in poverty, but represent 33% of the child welfare supervised placements and 25% of those in probation supervised group home situations. White children make up 39% of the general population and 24% of those in poverty (PIR 2006, 12). Nationwide, 14,300, or 17%, of our public schools reported 58,300 serious violent incidents. Serious violent incidents include rape, sexual battery other than rape, physical attach or fight with a weapon, threat of physical attack with a weapon, and robbery with or without a weapon (SSOCS 2008)

African Americans have the largest portion of its child, youth, and young adult populations in foster care ranging in ages from one to eighteen years of age (GAO 2007). African American families are less likely to have a father living in the home, which makes it more vital that close relatives learn how to help, placing children with relatives when the mother is found unfit. African American children are more likely than white and Asian children are to enter into the care of relatives. In 2004, African American children were found to be more likely to be diagnosed as having medical conditions or disabilities (28%) than white children in foster care (26%). Children with special needs required additional support services, and many African American families need this help to be more successful in taking on these mental and physical problems.

A school is a social place to learn how to learn to use academics to communicate using noises, sounds, signs, and symbols. "Children who are born to poor parents and live in poverty during their preschool years are vulnerable to certain kinds of disadvantages that keep them from making normal progress in their learning after entering school" (Costin, Bell, and Downs 1991). Many of these children come from homes where drug and alcohol abuse have led to inadequate nutrition, medical care, attention, love, safe play space, rest, and care. The conditions that surround a child's home, school, neighborhood, and workplace network affect how they learn, use language, and sense, feel, think, and react to people.

The Family and School Failure

Dropouts disproportionately make up higher percentages of the nation's prison and death row inmates. Approximately 34% of federal and state inmates and 51% of persons on death row lack a high school credential (U.S. Department of Justice 2004; 2007). Out of all public schools 83,000, there were 62,600 that participated in the survey, which accounts for 75.5% reporting 1,332,400, violent incidents throughout the primary, middle, and high school public systems. Violent incidents are defined as rape, sexual battery, physical attacks, fights, threats of physical attack, and robbery with or without weapons (SSOCS 2008). Our children are learning how to be violent in their schools, in the classroom, on school grounds, on the way home from school, on school buses, in school buildings, and at school sponsored events.

Learning how to go to school in the right state of mind has to be set up for the learner to experience. The school has to delve deeper into the reasons for acts of violence against children that come to school and are viewed as weak or different. Learning how learning with other people may begin in the home, but extends to the school, where officials must do their part to protect and serve children that come to them for help. The achievement gap between children of different racial and socioeconomic groups are tied to historical patterns of school failure in the United States. Hence, school violence is connected to race, class, and cultural experiences that influence student achievement.

The current learning and behavior of black children reflect those of a culture of children who methodically reject school and teachers (Ogbu 1978; GAO 2007). For example, black children have been found to be more likely to hit, curse, push peers, run about with no sense of control, employ disruptive, "silly" behaviors, not follow instructions or complete task, destroy property, and strike out at teachers (Taylor 1988; Slaton 2006). In a growing number of states, high school completion rates for African American and Latino students have returned to pre-1954 levels (Hammond 2004). One in every three black child is poor, and one in every four Latino child is poor. There are 32 million poor people in the United

States including 11.7 million children. Almost 75% of poor children live in poor working families. More than 5.1 million children live in extreme poverty (CDF 2003)

The Supreme Court's Brown decision, issued in 1954, dealt with the issue of access to education being limited by race; the Elementary and Secondary Education Act of 1965 dealt with the issue of discrimination based on economic class. The Education for All handicapped Children Act of 1975 dealt with the issue of inadequate services to children with special needs, and the No Child Left Behind Act of 2001 deals with the issue of how to live up to these historical promises (Woods 2004). In 2003, President Bush signed into law the Keeping Children and Families Safe Act, to reauthorize the Abandoned Infants Assistance Act under the Child Abuse Prevention and Treatment Act.

Hispanics make up 41% of the general population and 58% are in poverty. As of spring 2006, of the 47,000 students who fail to pass the California High School Exit Exam, 29% were English learners, 17% were low income, 19% were blacks, 18% were Hispanic, 5% were Asian, and 4% were white students (Public Advocates). Given the background and conditions of schooling for a significant number of black and Hispanic children, reflect issues of poor contact and interaction in larger numbers than in the pre-1954 period. According to the U.S. Census Bureau, 44% of youths ages 18 to 19 were high school dropouts. Nationwide, Hispanic males age 18 to 19 show the highest high school dropout rate at 57%. White males in the same 18 to 19 age group dropped out of high school at a rate of 47%. Asian males between the ages of 18 to 19 dropped out of high school at a rate of 41% (PIR 2006, 14).

According to Adelman and Taylor (2002), high-needs children are students who suffer from mental, emotional, or behavioral disorders, and relatively few receive mental health services. According to Marzano (2003), to deal with severe problems of learning and mental health, "Teachers require new strategies to advance their classroom management skills and acquire a repertoire of specific techniques for meeting the special interest of high needs children." According to Hammond (2004), "Special needs students, ESL new students, those with poor attendance, health, or family problems are increasingly likely to be excluded by being counseled out, transferred, expelled, or (pushed out) by dropping out."

The U.S. Department of Education has been lax about enforcing the No Child Left Behind Act reporting and accountability measures regarding graduation rates, while rigidly enforcing its testing accountability measures. An overemphasis on test-driven accountability, without the balance that graduation rate accountability provides, creates perverse incentives for school officials to "push out" low-performing students, and thus is likely to worsen the dropout crisis (the Civil Rights Project 2005). Nationally only about 68% of all students who enter the ninth grade will graduate "on time" with regular diplomas in the twelfth grade. While graduation rates for white students are 75%, only approximately half of black, Latino, and Native American students earn regular diplomas alongside their classmates. Graduation rates are even lower for black, Hispanic, and Native American males. Yet, because of misleading and inaccurate reporting of dropout and graduation rates, the public remains largely unaware of this educational and civil rights crisis.

Only 50% of black students and 53% of Latino students graduated from high school on time in 2001 (CDF 2007). The historian Lawrence Cremin (1990) reported on problems of school achievement as being relative to changes in the family, stating for one, "The family was the institution in which children had their earliest experiences with education, their earliest experiences in the learning of languages, the nurturance of cognitive, emotional, and motor competencies, the maintenance of interpersonal relationships, the internalization of values and the assignment of meaning to the world." Children who struggle with school represent a diverse culture of learners that have been hurt, which causes them to resist learning and performance goals. Every day, at least 2,261 young adult students, drop out of school (CDF 2007).

Children who do not experience academic or behavioral (social) success in their school years often become truant or drop out, and even if they finish school, they are inadequately prepared for life tasks and are assured underemployment, long periods of unemployment, and dependency on public welfare or other subsidies (Caple 1990). Children who fail in school report great anxiety and many worries, as well as being depressed and lonely.

The U.S. Department of Justice reported that about 1,498,800, of the prison population is made up of parents with minor children (PIR 2006, 11). Nearly half of those parents are black. Seventy percent of the state prisoners and 55% of the federal prisoners who are parents reported not having completed high school. Twelve percent reported that their education did not go beyond the eighth grade. Half of all Hispanic mothers, particularly immigrant Mexicans, have gaps in their basic skills and are mothers who do not have a high school diploma; and more than one-fifth of all African American mothers, particularly lower class blacks, have gaps in their basic skills and are mothers who do not have a high school diploma (p. 15).

Whites have the highest basic percentage achieving scores, followed by Asians, but Hispanics and blacks have different patterns of lower skills in basic literacy and quantitative areas. Mexican mothers have the lowest educational levels, with high school completion rates at 65%. Overall, Mexicans have the lowest high school completion rates at 70%, in comparison to other Hispanic ethnic groups. Asians, overall, have over a 90% high school completion rate. However, Southeast Asians completion rates have declined, in comparison to other Asian ethnic groups. Math proficiency tests gauge student performance in the area of quantitative achievement. Quantitative skills are important life skills and are required in technical and scientific fields. In fourth grade, over 80% of blacks and over 70% of Hispanic students scored below basic proficiency compared to about 40% of white and Asian students. In the eighth grade, black and Hispanic student's low basic proficiency scores remain high at 68% and 75% respectively.

The challenge is even greater for poor, minority, high-need, and special-need children in dysfunctional families, because of the many physical, mental, social, and emotional factors that can contribute to their poor interactions with his or her environment (Comer 2004). As we observe children move from the elementary grades to the middle school grades and to the ninth grade for high school, these problems become more apparent. "These challenges can overwhelm the coping skills of some students, lower self-esteem, and decrease motivation to learn" (Letrello and Miles 2003). Crises often develop between middle and high school, because his or her coping skills are no longer adequate in response to the deep academic focus required in high school.

The majority of California's 1,700,000, high school students simply are not reaching the academic levels needed to succeed in tomorrow's economy, in postsecondary education, or as effective citizens (CDE, (2005). About 30% of the entering ninth grade class fails to graduate on time. Research data suggest that the factors leading to student dropouts are in place by the time that he or she enters ninth grade. Despite decades of trying, research has not identified programs or services that consistently reduce dropout rates (Legislative Analyst's Office, (2005).

The gaps in academic achievement, dropout, and graduation rates for students from different racial and income groups are enormous in most school districts (Banks 2002). It is estimated by the CDE that of the 1,876, 927 youth enrolled for the ninth to twelfth grade in the 2003-2004 school year, 61,211 dropped out. In this sense, a family declines as a child moves through school falling behind.

Ethnic Group	Number of Dropouts	%
White Youth	13,043	21.31
Black Youth	9,938	16.24
American Indian or Alaskan	641	1.05
Asian Youth	2,461	4.02
Pacific Islander	528	.86
Filipino	943	1.54
Hispanic	32,610	53.27

Note: The table above reflects a competitive struggle to live and to learn in the home, school, neighborhood, and workplace settings, which establish social class and economic status in adulthood. Important social and economic factors are created in response to school achievement and school failure by each ethnic group, which leads to positions of dominance in mainstream society.

*Reflects problems living, learning, and participating
Source: CDE

Nationwide, white, Asian, and Pacific Islander youth are more likely to complete high school with a diploma than are their black and Hispanic peers. Over the last decade, 347,000 to 544,000, students dropped out of high school before completing the twelfth grade (U.S. Department of Education 2001. This data reflects that segments of our poor, minority, and high-need students are not able to sense, feel, and focus their brain in ways that allow them to participate in regular school events. Hence, they are less able to receive, structure, and express their feelings and thoughts in class. Both males and females are doing poorly. Nationwide, in 2007, the dropout rates of sixteen to twenty-four year olds was 8.7%: whites 5.3%, blacks 8.4, and Hispanics 21.4% (NCES 2009). The breakout for all males was 9.8%; 6%

for whites, 8% for blacks, and 24.7% for Hispanics. The breakout for all females was 7.7%; whites 4.5%, blacks 8.8%, and Hispanics 18%.

Graduation rates for minority males in California in 2002 reflect 50.2% for black, 54% for Hispanic, and 46% for Native American males. Black and Hispanic female graduation rates were at least 10% higher, at 60.2% and 64.9%. In contrast, graduation rates for white 74.6% and Asian and Pacific Islander 79.6% males were significantly higher; and the rates for white, Asian, and Pacific Islander were even higher for females, at 80.2% and 86.8%. Overall, Native Americans had the most alarming graduation rates at 52.2%, followed by blacks at 56.6%, Hispanics at 60.3%, whites at 77.8%, and Asian and Pacific Islanders at 83.5% by group (Swanson 2005). In other words, this means that Native Americans and blacks fall the farthest behind.

Extensive Human Systems Research has led us to identify being poor, minority, high need, and special need as the risk factors for school dropout and family decline. Of the 1,974,645 students enrolled for grades 9-12, 68,534, dropped out in the 2005-2006 school year (CDE, 2007). By race: blacks 165,683, 27% dropout; whites 688,641, 8.3% dropout; Hispanics 849,004, 19.1% dropout; and Asians 172,368, 5.7% dropout. By gender: Males drop out at a rate of 15.8% each year, and females drop out at a rate of 12.4% each year; males drop out at a rate of 15.8% each year, females drop out at a rate of 12.4% each year.

Race	Males	Females
Blacks	21.7% each year	21.7% each year
Whites	9.5% each year	7.1% each year
Hispanics	21.2% each year	16.8% each year
Asians	6.7% each year	4.6% each year

Note: the table above reflects the extreme levels of family decline, school failure, delinquency, and unemployment being experienced by children who live in poor; minority, high-need, and special-need high-risk home, school, neighborhood, and workplace networks.

* Reflects that a disproportionate number of black males and females will be pushed out of schools across the state.
Source: CDE, Educational Demographic Unit. 2007. Statewide Dropout Data.

Learning Disabilities

According to the National Center for Learning Disabilities (2003), the number of children classified as having learning disabilities has substantially increased in the last twenty years. Since the 1980s, significant mental defects emerge as children approach school age because their parents used drugs through the time of pregnancy. The U.S. Department of Health and Human Services reports, "The American family, 8.3 million children in the United States, approximately 11%, live with at least one parent who is in need of treatment for alcohol or drug dependency. Children of parents who abuse drugs are three times more likely to grow up being verbally, physically, or sexually abused, and are four times as likely to be neglected as other children are."

When a child has an academic disorder, it may take years for them to be diagnosed. Academic records must exist, which shows signs of academic failure (poor grades) over a period of years. A parent has to prove these records reflect low reading, writing, math, and speech scores. According to the National Institute for Mental Health, "In fact, reading disabilities affect 2% to 8% of elementary school children" (1999, 4). The skill to write is affected by a child's skill to read, which is affected by his or her skill to speak and listen. For example, writing involves several mental, physical, and social brain functions. The brain networks for vocabulary, grammar, hand movement, and memory must all be in good working order (p.5). When doing math, the child must be able to recognize numbers and symbols, memorize facts, multiplication tables, align numbers, understand abstract concepts such as place values and fractions.

The poor performance and achievement problems of black males as a group has been related to internal factors that include issues of self-concept and identity (Tatum 2006). "Black/African American students are less likely to come from two-parent homes than Asian-American, Latino, or white students"; the stigma associated with being racially marked as "black" undermines the opportunities for African Americans to fulfill their potential as individuals and as a group, and the racial stigma of being black influences patterns of contact and the various mechanisms (ethnic class differences) that have bearing on social mobility cause them to isolate or be isolated (Rowley 2004). This fact alone, in terms of educational quality, suggests on these grounds that black students are more likely to experience social disorder along with Hispanic students involving violence and lower quality instruction within their home, school, and neighborhood networks.

Factors related to drop out of students with learning disabilities include low socioeconomic status of family, lack of books and other reading materials in the home, level of schooling of the parent of same gender, low grades in school, teen pregnancy, prior academic failure, and the use of cigarettes, marijuana, and other illicit drugs, along with aggressive behavior; absenteeism, course failure, and peer influences (Dunn, Chamber, and Rabren 2004). Many students with disabilities experience discrimination or experience inadequate educational programs, because their racial, ethnic, social class or gender is different from the majority (Banks and Banks 2003). Black students comprise about 16.3% of the general school population but more than 31% of students classified with mild mental retardation and 23.7% of students with severe emotional disturbance.

The study of delinquency sets the stage for learning how the home, school, neighborhood, and workplace network unfold through the minds of helpers, and the bodies of growing children, youths, and young adults. For example, their lack of information on how to help, or how to accept help relates to the knowledge they have or lack thereof, because in the pursuit of the experience or the lack thereof, they are left with less to reflect on in the study of the human system. Delinquency begins with a child, youth, or young adult's responses to family, education, government, and business practices and unfolds with each human-to-human contact, from one stage to next. This process gives rise to his or her state of mind and use of body language. Hence, the emphasis is on the act or behavior in the form of a resistant state of mind to being led, controlled, or managed in one or more of these fields.

Let us be clear. The home is linked to family, the school is linked to education, the neighborhood is linked to government, and the workplace is linked to business fields. Within these linkages the individual stands for the self, the group stands for other people, and the environment of contact stands for the home, school, neighborhood, and workplace network as a human systems study.

Human Systems Research

We know how to use information to move parents and children in crisis through their home, school, neighborhood, and workplace networks.

We estimate that about 80% of the children youths, and young adult that have come through the Family Leadership Academy since 2000, have graduated from high school. Family-centered learning allows us to inform and instruct parents and children while we use the information to improve human relations through community-based learning practices, products, and services. We then deliver an ecological perspective, which symbolizes the home, school, neighborhood, and workplace network of human beings and their affect on health (Hilgenkam and Kathryne 2006; Slaton 2008).

The term *home* refers to the basic set up for sustaining life in a man-made ecological system. The term *school* refers to the domain of mechanics and the socialization of human experience. The term *neighborhood* refers to the people and environment, which contains the conditions for sustaining health and safety. The term *workplace* refers to the conceptual lens and images of the people living, learning, thinking, and responding to the underlying principles of the economy as the human ecology. The HSR approach allows us to set up a scientific, child-centered, progressive, and transformative view of the family as a business. This system works for children, youths, and young adults in crisis, because they need to learn how to live in a home with a sense of family values, learn in a school with a sense of educational values, think in a neighborhood, with a sense of government values, and respond in a workplace with a sense of business values.

HSR is the type or study that seeks to integrate both natural and mechanical systems. This leads to emerging forms of information, knowledge, experience, and reflection that are both inclusive and progressive (Jeffrey 1980). We use a highly structured family-centered learning model and mixed methods: action research (Carr and Kemis 1986), interpretive research (Denzin 1989, 1997), community-based action research (Stringer 1999), and HSR (Slaton 2005). With these perspectives in mind, we strongly believe, parents, teachers, and principals need a rigorous home-to-school process designed to equip all students with the capacity to practice what they learn, to produce a broad base of knowledge, cognitive skills, and attitudes suitable to service learning, which requires cooperative thinking, creative problem solving, and give-and-take flexibility.

Poor, minority, and high-need children, youths and young adults that do not respond well to the mechanics of going to school in the right state of mind represent the human factors—influence on health, safety, and the quality of well-being, because of the biological, physical, chemical, social, and psychological, human behavior of being more resistant to school. Merton, Reader, and Kendall (1957) described socialization as a process to which parents, children, teachers, and administrators must make and sustain their contact long enough to forge effective contacts to help acquire a sense for and to help share the values meant to forge compliance. The ecological or human systems model considers the whole interaction processes that constitute parent, child, teacher, and principal contact, which creates the context for, and sustains, the divisions of labor between them (Slaton 2008, 4).

The role of the parent extends from the home, the role of the child extends from the parents leadership in the home, the role of the teacher extends from the classroom to the school, and the role of the principal extends from the administrative office to the school. Parents parent, children study, teachers teach, and principals manage these relationships. Everyone experiences the contact in their home, school, neighborhood, and workplace network. These are the events that fuse how we live, learn, think, and respond in relation to each other.

The child, youth, or young adult is hurt when they cannot live free of their emotion to self-destruct at home, at school, in the neighborhood, or in reference to a focused state of the workplace, as indoor or outdoor learning in the context of being controlled. Most teachers and principals see the roles and functions of teaching and learning as complex activities, which are intertwined with what children bring to school. Hence, they think of schools as places where children become literate and well behaved, learn to display solid character, and become connected to the communities. On the other hand many corporate leaders see schools as places to produce a product, such as academic achievement, and sorely in need of marketplace moxie (Cuban 2004). This suggests that the economy is the underlying human science, which further implies that ecology forms the basis for human learning, that learning about the economy is controlled.

For instance Cuban (2004) wrote from a transformational perspective on the industrial age thoughts of renowned social reformer Jane Adams. "The business man has, of course, not said to himself: 'I will have the public school train office boys and clerks for me, so that I may have them cheap'; but he has thought, and sometimes said, teach the children to write legibly, and figure accurately and quickly; to acquire the habits of punctuality and orders; to be prompt to obey, and not question why; and you will fit them to make their way in the world as I have made mine" (p.

1). Today, this is a profound statement. It shows a lack of consideration for the child's cognitive capacities and potentials, which have led many children, youths, and young adults of this day to question the value of school-based learning. In today's world, we are concerned about the effects of major life events that are ignored—being born to mothers who used street drugs while the child was in utero or after giving birth and on; the experience of being abandoned, neglected, abused; and living with little or no leadership and/or structured development.

We explained why research-based special education services were needed in a March 8, 2006, evidence-based report to State Superintendent Jack O'Connell. The research-based approaches we propose seek to learn how to move children, youths, and young adults through the school system, more effectively than what is now being deployed to achieve this aim. Our goal is to rescue and help children that have been hurt or injured by major life events, which affect major cognitive structures in how they grow, mature and develop as human beings because of poor health, social and academic enrichment, all of this, causes it to be more difficult to live and learn; especially in learning environments without special support and educational services (Slaton 2006, xvi). The improving America's School Act of 1994 is a reauthorization of the Elementary and Secondary Education Act of 1965, which is extended through the No Child Left Behind Act of 2002, and the Education Sciences Act of 2002, for scientifically, based research in education.

To increase the scientific knowledge base and to develop educational programs, products, services, and practices that will enhance student outcomes for disadvantaged children and youths attending our public schools in the twenty-first century, knowledge about him or her as student participant in education is required. Scientifically based qualitative research for observation and experimental purposes provide the best approach to develop meaning and understanding of the home to school context, where causal factors emerge, because of natural human activity. Using interviews with key persons in the role to educate and develop the lives of parents and children experiencing the crisis of disconnect, because of poor home to school learning, we seek to learn why to form, organize, and construct meaningful and interpretive reports that address the causes of family decline and school failure.

Community-based action research is a research plan that focuses on ways to develop patterns of participation through facilitation that studies the teaching and learning process for the best practice to inform and instruct less focused learners as student participants. The guiding theory of the design is to learn how to teach each participant to improve contact, interaction, and cooperative levels of participating in home and school events long enough to make sense of the experience. Progressive Investing Workshops are used to assess the effectiveness of clinical trials and in respect to achievement (improving) test scores. The quantity of evidence is to be set up based on family, medical, and educational histories. Progressive Investing Workshops are to be employed as the experiential base for quantifiable knowledge through participant surveys, participant evaluations, and participant testimonials.

Qualitative evidence, such as changes in attitude, behavior, grades and test scores and aggression: The home, school, neighborhood, and workplace relations of the participant's family and educational dynamics are studied to learn how they learn in these places, from where they are, as people in crisis. The parent and child's learning system is set up to help them practice the act of participating in self-development, information processing and community-based learning tasks. This allows us to deliberately link social and academic learning to how a family learns to use education alone and with other people to live. The relevance of this evidence becomes clearer in reference to the need to improve the ability to learn at the elementary school level, the need to improve the ability to be a learner at the middle school level, and the need to improve the ability to be a student at the high school level.

Because of the magnitude of the need for this evidence, it is expected to reduce the costs of education, delinquency, unemployment, and family decline. The context, which contains the causal processes for maintaining anger, violence, and hate, will produce better student outcomes to organize, to plan, to guide, and to implement prevention and intervention practices to change the set of events hurting or injuring the family and educational life span. Organizing a child's medical, family, and educational history does this, which promotes scrutiny and criticism of the changes, as the child moves from elementary, to middle school, and on through the high school grades. Thus producing a rise in small learning communities of parents, teachers, and children (students) too form and construct new attachments to improve their home to school knowledge base, while working to fulfill the basic task in becoming a concerned citizen and participants in their own success. Hence, the child, youth, or young adult's right to the golden opportunities of citizenship are better protected and placed within his or her reach while in grade school.

Our goal is to provide supportive educational services to youth that have been hurt or injured by major life events, which causes them to disconnect from home to school learning, before they have acquired enough social and academic skills sets to cope with life after grade school. Our goal is to provide supportive educational services to young adults that come from a low-achieving family, because of poor school contact, interaction, and cooperation in learning how to be a student and participant. These goals coincide with the procedures to be used for identifying the characteristics of the research study to be included in evidence-based reports to educational leaders. Community-based action research is an inclusive research design that is sensitive to the realities of research in theory and in practice (Slaton 2008).

In our Building Human Asset Meeting with Deputy Superintendent Anthony Monreal, CDE, Dr. Monreal asked, "What is the next step?" Our reply was in reference to the information we have requested in accordance with the Posny report (from the U.S. Department of Education; from the Office of Congresswoman Doris Matsui, May 10, 2007) and the crisis facing our youth in public schools across the state. School failure of this magnitude can only be addressed through the scientific development of new practices, products, and new services to improve the learning, behavior, and practices of our youth, parents, teachers, and school principals' human relations.

We reported to Dr. Monreal, who is responsible for the Curriculum and Instruction Branch and Deputy Superintendent Rick Miller, who is responsible for Policy Development and External Affairs. We are in talks and negotiations with several school districts that do not appear to be acting in good faith. Dr. Monreal advised us not to stop negotiating with those school districts and to email him concerning an event where he could learn more about the Human Systems Research in Education pilot project we propose as a recovery plan with statewide implications. This was all smoke and mirrors. We were being set up to be deprived of our intellectual property.

"All school districts have the authority to negotiate contracts for almost everything from food services to janitorial services to security services to collective bargaining agreements with teachers. Large school districts may regularly maintain as many as 150 contracts for a complete range of services. School districts also have the authority to contract for educational and instruction services. Often schools will contract for special education services, media services, even for athletic coaches;

often they contract for entire alternative schools" (Barr and Parrett 1995). It is not unusual, for instance, for a number of schools to be asked to work together collectively and contract with a single service provider at the state and county level to provide support services that would serve students from each of the member school districts.

As we reported to Governor Schwarzenegger and Senator Steinberg, Save Our Youth the Next Generations (SOY) expressed its concerns about the free and appropriate public education (FAPE) provided to poor, minority, high-need, and special-need children to Congresswoman Doris Matsui. The questions posed by SOY were passed on to the United States Department of Education, Office of Special Education and Rehabilitative Services. Their response to SOY addressed four questions posed by Congresswoman Doris Matsui, based primarily on the Individuals with Disabilities Education Act (IDEA). The four questions posed by Congresswoman Matsui to the U.S. Department of Education on Save Our Youth's behalf were linked to federal law. Part B of IDEA provides guidance to the States and through them to local school districts. These four questions were intended to guide the discovery process, inquiry, and request for assistance in obtaining this information. These questions were as follows:

1. What criterion is established for determining eligibility status for special education classes and vocational rehabilitation services?
2. Which entities are accountable for determining the assessment and methodology for special education cases?
3. Is there a current standard for addressing children who come from a family suffering from substance abuse (related disorders)?
4. Are family history and medical records taken into consideration when assessing a child's eligibility?

This information was crucial to our ongoing campaign to improve learning and support services that will enable poor, minority, high-need, and special-need students caught up in poor home, school, neighborhood, and workplace networks to learn more effectively and become contributing members of society.

We are concerned with "why" so many poor, minority, high-need, and special-need males between the ages of twelve and twenty-five are entering the California Department of Corrections and Rehabilitation

(CDCR) as delinquents, dropouts, and criminals with learning and mental health problems that were not being adequately assessed by the CDE. This is where the "social problems" of family decline, illiteracy, recidivism, and poverty emerge as a result of having been overly exposed to poor home, school, neighborhood, and workplace networks that harbor fear, anger, and anxiety toward poor, minority, high-need, and special-need children and families that are caught up in them; and hence, made more dysfunctional by the mere magnitude of these problems within the cultural context of family systems.

We are concerned with "how" poor, minority, high-need, and special-need children, youths, and young adults need to be educated when they have been hurt by major life events that have a significant impact on their learning, behavior, and mental health. With recent state changes in the Division of Juvenile Justice within the California Department of Corrections and Rehabilitation, more children, youth, and young adults are being incarcerated to provide the treatment and training that would be impossible to deliver in regular home, school, neighborhood, and workplace networks. One of the purposes for this concern is because too many children, youths, and young adults are moving through the public school system with learning, behavior, and mental health problems that are not being assessed by professionals in medicine, education, corrections, and mental health in a timely manner to determine whether or not a FAPE is actually being provided.

Research has documented repeatedly over the history of the public school that black and Hispanic students do not receive the same quality of attention, time, and care as do white students. Hence, parents, teachers, principals, and students who choose to participate in home, school, neighborhood, and workplace improvement tend to get along better when there is a structure in place to help them each advance their part in public education. Only 38% of black students successfully passed the math portion of the exit exam, compared to 70% of whites. Forty-nine percent successfully passed the English or language portion of the exam, compared to 74% for whites and 50% of Latinos (SABC 2007). In a large way, school interaction problems acts as root causes for why children, youth, and young adult students do not relate well to the people in a particular school setting.

As a reform strategy for transforming parent and child responses to their schools, we devoted two years to the study of human relations issues in the Del Paso Heights Elementary School District (DPHESD),

Elk Grove Unified School District (EGUSD), and the Sacramento City Unified School District (SCUSD. We have met school officials from each of these districts through the consent, and in most instances, direct meetings with superintendents to discuss this approach. For us, the natural next step was to request their participation in a BHAM to share these experiences with both deputy superintendents Monreal and Miller.

Hence, the focus of our February 29, 2008, BHAM was on how to talk to the child's brain, to inform civic leaders, educators, educational policymakers, and stakeholders through HSR model. By the end of the meeting, parents and children from several school districts had taken advantage of their opportunity to share their stories. Board of Supervisor Don Notolli stated on the record, he stood in support of our work with children and families in crisis. Members from the California Department of Education (CDE) said on the record, they were willing to work with us to establish some funding sources. A local school district pledge to look at more ways we could partner. These were promises made, but never kept.

As we reviewed the video footage from our BHAM, we could sense from the language being used by our attendees, they were taken back by our levels of expertise and practical knowledge of the education system. The experiences that were shared revealed that school leaders must start thinking through how they plan to work with children, youths, and young adults that come to them hurt by major life events. The system is not designed to respond to children in crisis or students with learning problems, behavior issues, and conduct concerns—which means we had to think of ways to lead them to review HSR.

We asked Gordon Jackson, the division director of the Curriculum Instruction Branch of the CDE, "If there is another organization doing HSR in the State of California to please let us know." Mr. Jackson said the reason for his request for more information about us was to determine if the services we were offering already existed in the system. We wanted to know if there were other competitors, and whether their work related to the systemic improvement of the home, school, neighborhood, and workplace networks of children and parents in crisis due to pandemic circumstances. "The drug epidemic, leading to 'drug babies' and high rates of children being born addicted to drugs: Our proposal to the EGUSD specifically addresses this crisis in a Special Services Proposal" (Slaton 2008).

In our meeting with the chancellor of the Los Rios Community College District (2007), we presented a strategic action plan for rescuing high-need young adults with learning, behavior, and conduct symptoms that affected their attendance and retention. We went on to meet with the president of Sacramento City College, the vice president, and the vice president again, along with four deans from various departments. A human learning center at the junior college level for young adults with substance related injuries or hurt from problems in the home, school, neighborhood, and workplace networks would improve campus safety. We explained at each level, "This is an issue of safety, because too many children youths, and young adults are displaying learning, thinking, and behavior problems that cause them to withdraw and resist the help of helpers (Slaton 2009).

A cursory look at the subjects of our Progressive Investment Reports to federal, state, county, and local officials puts into perspective the nature of this resistance. We have submitted the following reports:

1. 1998-2004 Progressive Investment Report: SOY Family Leadership Academy, Building Human Assets Process for Attacking Urban Decay, Research-Based Practices.
2. 2005 Progressive Investment Report: Research is ongoing, but what is shared in this report was shared with school officials; SOY Family Leadership Academy; Research-Based Practices.
3. 2006 Progressive Investment Report: Evidence-Based Reports to our Public Schools; why many poor, minority, high-need, and special-need children cannot learn.
   a. Request for Building Human Asset Meeting to Discuss our Request for Investigation.
4. 2007-2008 Progressive Investment Report: Help to improve America's schools
   a. Search for federal or state policy to address the eligibility status of children with a family history of suffering from substance abuse related disorders.

On July 7, 2008, we were contacted by Cris Forsyth, chief of staff for California State Assembly Member Jim Beall Jr. in response to our request for a BHAM. Mr. Forsyth wrote, "As Chair of the Assembly Select Committee on alcohol and Drug Abuse and the Assembly Committee on Human Services, he is committed to addressing the issues that concern your organization. This is great news, we thought. As we informed Cris Forsyth: 'We use Building Human Asset Meetings to inform and instruct our leaders concerning the work we are doing and to learn how to obtain the information we are requesting from the CDE to improve learning and support services to poor, minority, high-need and special-need children and youths that are growing up hurt by major life events in their home, school, neighborhood, and workplace networks. We would like to discuss these issues with Assembly Member Jim Beall Jr."

In reflecting on these meetings (3) with Assembly Member Jim Beall Jr., I am reminded of how he informed me; his colleague's interests are far removed from our cause. At this point, we began to talk about why we may share personal interests in review of the resistance to our advocacy for poor, minority, high-need, and special-need children; children with substance related disorders; children in need of improved learning and support services. He was the first and only political official to just plainly say they do not care. In 2008, we wrote to every California legislator requesting a BHAM to obtain their support. And the assembly members words began to appear true, as all those that had pledged their support for our cause in 2008, disconnected from us without so much as a phone call, email or follow up BHAM.

The United States Constitution provides indirectly for the federal government to exercise a role in the three-way partnership of government responsibility for education (Brimley and Garfield 2002). Chris Flores from the office of U.S. Congresswomen Doris Matsui referred us to the office of Congressmen George Miller, Seventh District, and California. We met with Ms. Barbara Johnson on October 7, 2008. She accepted a copy of our 2007-2008 Progressive Investment Report and said the congressman as the chair of the Education Committee would review it and respond to us in writing. Part of the problem in considering the philosophy of behaviorism is that the education system works under the guise that cognitive experiences and mental processes do not make clear, how the child learns.

We are describing a problem with the system of public education. Knowledge of what it means to be human, what it means to have cognitive capacity and potential, and what it means to have a behavior pattern is necessary to understand how any child learns, but more to our point, how a child that has been hurt by major life events learn. All three are basic elements in how humans learn from certain experiences to set up behavioral controls to reduce the effects of emotion. The system of public education has to move into the information processing age; this is the evolutionary process. The product of which is knowledge and experience.

Review the models we have presented above. Knowing how to live and wanting to learn how is complexly intertwined.

A BHAM can be used to inform, instruct, or motivate. Our language system is related to the cause, or the responses of all the officials that have met with us to learn more about human systems science. For instance, we met Mayor Kevin Johnson, of Sacramento, at one of his community-based office hours. He advised us to contact local council members to discuss our proposals. We received another call inviting us to another office hours where the mayor stated, "I share your concerns for our children and families." We were asked to contact Lyn Corbett, director of the Office of Youth Development to submit a proposal. We met with Lyn Corbett on January 6, 2009. Knowing about human systems science is one thing, but experiencing it firsthand, the message is received and better understood, Mr. Corbett stated.

Mr. Corbett pledged that he was partnering with us on that day, January 6, 2010. But when we called to follow up, Mr. Corbett was not accepting our calls, our e-mails were being rejected, and then finally, we requested that our materials be returned. Mr. Corbett wrote, "It was a pleasure meeting you. I look forward to working with you in the future" (March 1, 2010). We know our message got through. To us, his response that said "How do you know what you know?" came through as well. Hence, if nothing else, we knew our science was sound. As in many other respects, receipt of the message: Where are your signs of care? Receipt of the message can be viewed here as comprehended based on sense data, but resisted because of what changes in behavior would look like. As noted, the process of achieving change requires new learning. Recall that we are asking our public officials to move from a state of knowing to a state of the unknown.

Hence, the resistance to us is because change entails having to learn from us. But California Assembly Member Alyson Huber's resistance to us was more blatant. We met with Assembly Member Huber's office on December 1, 2009, Mandi Bailache, the assembly member's assistant, to discuss the education of children and parents in crisis. The meeting took place in the reception area; I was accused of being disrespectful for questioning the location and readiness of her legislative assistant by the chief of staff. We filed a complaint with Assembly Member Huber's office on April 14, 2010. What was most offensive to us were the March 3, 2010, and April 10, 2010, letters we received from the Assembly Members office.

The first one is written to thank us for sending her a copy of our report. And to let us know "As legislators, we depend upon the expertise of the state Department of Education [DOE] to examine new proposals as they are developed. I understand that the DOE has met with you to review your program, and our state Superintendant of Schools, Jack O'Connell has also been involved in this process." Now this was kicking the can down the road. "By the way, I do have a copy of your book and I am familiar with your approach to education." One month later we received the second letter noted above from Assembly Member Huber's Office.

"As a mother of two boys, I understand that choosing child care is an important decision and every parent wants to make the right choice for their child. As your representative in the State Assembly, I want to make sure you know about some of the resources available when choosing a Licensed Family Day Care Home or a Child Day Care Center." As a constituent and business person in Ms. Huber's District, she failed to show us any signs of care. "Unfortunately, because I disagree with the current policy and procedures used by the CDE to screen and assess children for traumatic injuries, behavior issues, learning problems, conduct concerns, and emotional disturbances our request for participation are being met with high levels of resistance" (April, 14, 2010).

It is difficult to imagine how she really feels about our book *Education and Science* (2009) because she did not respond. However, she did let us know that she has a copy of our book and is familiar with the approach we advocate for to improve the delivery of learning and support services. This was shameful. In an April 21, 2010, report to the state superintendent of school, Jack O'Connell, we filed a complaint against what we see as shameful and bias practices.

We asked the superintendant not to continue to ignore and reject the value, validity, and relevance of our work with children and families in crisis. We asked for the official responses from the DOE to our proposals being referred to by Assembly Member Huber in her March 3, 2010, letter to us. Unfortunately, this has had a negative effect on both members of the Senate and Assembly, whom we have asked to review our proposals for cause. Hence, because of this, we have been met with high levels of intolerance by the CDE. And this has made it more difficult to persuade more representatives of the state of California to attend BHAMs to learn for themselves the value, validity, and relevance of our work with children and families in crisis (April 21, 2010).

Poor, Minority, High-Need, and Special-Need Males

The Building Human Asset Project for improving learning and support services focused on children, youths, and young adults moving through the public school system, with signs of learning, behavior, or conduct problems. A learning problem is one that impacts a child's capacity or potential to receive, process, or respond to academics. A behavior is a response to contact that impedes a child's capacity or potential to sense, feel, or focus their brain and body activities. A conduct concern is one that affects a child's capacity or potential to choose, comprehend, or change a poor behavior pattern due to a lack of internal control.

Larry, now a seventh grader in middle school, has met with an IEP team that has decided additional assessments are needed. Though he is failing in all subject areas, Larry has a positive attitude and works well with teachers but cannot transfer what he hears to print. As a young child, he began experiencing difficulties with his use of talk, and he has never been a successful student. His father is in prison and his mother is hooked on street drugs. His mother is ashamed not only of her son's poor grades in school, but of his lack of toughness in their neighborhood as well. Larry's grandparent simply reaches out to him hoping he will keep "trying."

When we met with Larry's mother—to learn the family history—her talk revealed a pattern of male decline, which dates back to the 1970s where no male children have graduated from public school with a high school diploma. Each generation has been caught up in either street drugs, gangs, or both. In Larry's case, we learned from our talks with his grandmother and aunt that he had been subjected to drugs while in utero. Larry's father, also a drug user, was violent and abusive. Larry is being pulled into the street life by his dad's gang members, and the fact he is not doing well in school, helps to push him out of school.

Larry is growing up hurt by multiple major life events that impact his capacity and potential to learn how to receive, process, and respond to academics. He has a behavior pattern that shows signs of feeling hurt to the extent that he does not seem to care about school. For instance, when he uses talk, he curses without signs of awareness of where he is and who can hear him. He is constantly being teased and bullied by peers and is becoming more riddled with signs of anger. He has begun to lie to his mother and steal things from his grandmother. And he does not feel safe at home or in his neighborhood.

Despite all this, his English teacher reported him as being a good kid. His grandmother and aunt asked us to help. The school did not. His mother did, but she is moving deeper into a life of street drugs and binges that make her unreliable as the parent. What did his neighborhood look like? We had to visit his neighborhood to learn more about the pressure he was experiencing, as well as the obvious stress he was feeling from his parent. The task for human systems researchers is to find the signs of care in Larry's home, school, neighborhood, and workplace networks that may be set up to help him, help us.

These are the informal steps that need to be taken by human systems researchers to create a family-centered learning file. We use this intake process to study the circumstances in a child, youth, or young adults home, school, neighborhood, and workplace network to learn through the senses. Who are the best people to meet with in this good kid's problem situation? We want to learn if they are open to learn from us and with us in the lead. Hence, to learn more about the circumstances, we look for elders and try to work back.

This way, a family-centered learning file will contain data on the authentic core of competence in a group of people. The file is then handed off to a human learning consultant. The human learning consultant uses the file to meet with the parent and child to set up a series of workshops. At this point, the family-centered learning proposal (FCLP) is designed and individualized based on what was learned by human systems researchers in their home, school, neighborhood, and workplace visits. The human learning consultant assesses how the parent and child use talk to interact. What things do they have in common? Who is hurt? Where is the hurt coming from? Do they show signs of care?

The area school is notified that the parent and child have asked us for help. We ask the parent to fill out an information release form, and we ask the child if it would be okay for us to visit the school. If they agree to both requests, we then ask the parent to write a note to the school that authorizes us to assess the child's contact and interaction in school. Building Human Asset Meetings are set up with the parent, teacher, school principal, and designated staff. We use about twelve hours to develop these core assessments. What we look to learn is how the school people relate to the parent and child's dilemma and how the parent and child respond to the teacher and the principals contact.

By the time a student study team, IEP team, parent-and-teacher conference, or parent-and-school principal meetings are set up, the parent and child have participated in our workshops. Hence, as we meet the people in the child and parents network, the FCLP is being produced as the parent and child moves through a minimum of twenty-four hours of workshop events. This way, the parent and child are allowed to choose to participate, comprehend their need to participate, and change any course of action not approved in their FCLP.

Service learning projects are set up to allow the parent and child to display the effects of their workshop participation in a formal setting. They tests how feelings are sent forward and how to capture feelings received backward. In other words, the parent and child are more aware of the words we use to inform and instruct them concerning their contact, interaction, and levels of cooperation needed to participate in the act to move through the event. The student's school records are reevaluated to learn if there is sufficient progress being made.

To help parents, children, teachers, and school principals learn in relation to the needs of each other that they have been offered the BHA model. In doing so, we have felt the need to call for a strategic task force to study drug-related disorders and the growth in children with violent tendencies. We can use this class of human systems research to systemically improve the learning and support services of poor, minority, high-need, and special-need children. The more we meet with parents and children to learn about the major life events that affect their ability to live and learn how to lead and follow, the more we learn that our public schools are not prepared to inform and instruct children that come to them in crisis. We have known since the late 1960s that street drugs posed the most deadly effect on childhood characteristics that produce responsible young adults.

The human learning consultant is a strategic thinker in the area of human systems research. To learn how poor, minority, high-need, and special-need students learn and behave with problems, we study the way he or she may need to be set up in home, school, and neighborhood events as the sources, which informs us. Think about the problem children who were born to mothers who used street drugs since the 1980s face in our comprehensive schools. Remember how you may have first learned about drug babies and heard doctors and nurses make those first reports—always crying, in need of endless attention, and frequently being abandoned.

The number of babies born to mothers who used street drugs before and during the time of pregnancy has grown dramatically over the past thirty years, and it appears that this number will continue to increase. Between 1976 and 2006, the overall number of school age children with learning and behavior problems has increased, along with the rising use of foster care and group homes for neglected or delinquent children. Hence, all through this same time, the number of children who reach school unprepared for contact with other children and school officials has grown as well. Altogether, these children can be identified by his or her lack of reading, writing, and math skills in structured activities.

We now know more about the problems these children face in relation to being teased and bullied, which increase learning and behavior problems due in part to poor school experiences. If they are under attack, then you are under attack. If the homes and neighborhoods they live in are under attack, then the schools and workplaces they need to live are under attack. They are in need of relief, safety, guidance, learning, and support. You are in need of a strategic action plan to reduce these threats on school officials and their clients. We are strategic thinkers with a clear record of success in this crisis.

In addition, learning and support services have always been the standard of improvement. Parents, children, teachers, and school principals rely on strategic advances in learning and support services to create and sustain successful contact with poor, minority, high-need, and special-need students. We have presented concrete stories, produced clear explanations, and have constructed lucid reports on how to help improve the reading, writing, and math scores—all through a focus on how they learn in home, school, and neighborhood situations. Many of these children, youths, and young adults run out of time. Today, these children represent an important segment of our population—that age group between infancy and age thirty.

With the influx of drug-related disorders in mainstream homes, schools, and neighborhoods today, this also means that an equally significant segment of our adult population is not prepared to discuss this issue. What we in mainstream society have not had to think about in the past, we now have to consider with care, as well as with concern for our current and future generations of school age children. We are experiencing the highest levels of home-to-school violence in terms of learning, behavior, and poor responses to contact. We can associate these responses to the drug epidemic, because too many parents, children, teachers, and school principals stand divided along these lines and in strict reference to the use of special education.

In the DSM-IV-TR (2002), substance-related disorders are caused by the effects of the drug abusers addiction and behavior (substance-induced disorders have an effect on a child's social, cognitive, emotional, and behavioral growth). For instance, early signs of child mental illness due to cognitive, behavioral, or physiological symptoms are similar to patterns of withdrawal in adults. Specific cognitive impairment, such as lack of ability to focus and control mind and body activity, and specific emotional distress, such as signs of anger, fear, and anxiety through acting out, and specific behavior problems, such as signs of resistance, poor social reactions, and disconnection. The maladaptive patterns of learning in social situations help to create unhealthy tension in parent, child, teacher, and school principal relations.

It is very important to us, as advocates, that we respond to requests for help. More specifically, to the parent and child who chooses to participate in workshops to learn how to comprehend their need for change. This is important for two reasons. First, the U.S. Congress passed the Individuals with Disabilities Education Improvement Act in 2004. Only by collecting research-based data can this mandate improve what we know about parent, child, teacher, and school principal interaction. The nature of the data we collect allow us to study how open or closed people are in roles of leadership, to this study. So far, only a few leaders have shown signs of care. For example, parents have spent more than 208 hours trying to understand the basics in how to advocate for children with early relationship problems.

This is expressed through our concerns about the class of a free and appropriate public education (FAPE) being provided to poor, minority, high-need, and special-need children to school officials. The questions posed are passed on to the groups of people we hold BHAMs with to learn how to help. Their response to us is related to the four questions posed by Congresswoman Doris Matsui, based primarily on the Individuals with Disabilities Education Act (IDEA). This is a federal program administered by the Office of Special Education and Rehabilitative Services and includes some of the requirements of the Vocational Rehabilitation (VR) Act. The Rehabilitation Services Administration, another component of the Office of Special Education and Rehabilitative Services (OSERS), administers the VR Act.

Save Our Youth's concerns are based on the lack of structure. Poor, minority, high-need, and special-need children, youths, and young adults should be able to matriculate through the public school system and receive all of the services they are entitled to under FAPE before their eligibility runs out. Hence, experts—professionals from the fields of medicine, education, corrections, and mental health-ought to be arranged in such a sequence of events as to determine the most effective method to improve the learning, behavior, and mental health support services to the child and parent. If taken into serious consideration, the assessments of children, youths, and young adults from the ranks of poor, minority, high-need, and special groups would be implemented as a safeguard against the crisis of family decline, illiteracy, recidivism, and poverty as early as the preschool stage. This would allow the learning and support services to be identified at earlier stages of his or her cognitive, behavioral, and developmental phases. Hence, this would greatly reduce the additional costs of educating, housing, and providing these services later on in the service delivery system.

We have met with the Office of Congresswoman Doris Matsui, the chancellor of the Los Rios Community College District; the CDE deputy superintendent Sue Stickel; the County Board of Supervisor, Don Nottoli; the president of the California NAACP, Alice Huffman; the president of the California Black Chamber of Commerce, Aubrey Stone; and many other outstanding leaders. These BHAMs were requests for assistance as we asked each one of these distinguished leaders to help us learn more about the problem of having so many males come through the public school system and reach adulthood lacking any signs of being able to learn, behave, and control their mental states of mind in home, school, neighborhood or workplace settings. In other words, we discussed how, why, and who these males were that make up 95% of the California Department of Corrections and Rehabilitation youth populations and the impact this problem has on the home, school, neighborhood, and workplace networks of children and families in crisis.

We need the parent and child to ask us for help so that we can ask our leaders for help.

1. Because the number of children with a parent in federal or state corrections facilities has increased by more than 100% since 1991.
2. Because real and measurable progress has not been made at any point in standards-based education for males.
3. Because the cost to house one youth has reached $200,000 (CDCR 2007).

4. Because these are male youths by ethnicity aged 12 to 25: Hispanic 51%, African American 31%, white 13%, and Asian 2%. Statewide, the gap between poor, minority, and high-need students continues to increase. Proficiency rates in English, language, and arts: Hispanic 25%, African American 22%, white 39%, Asian 45%, and Filipino 47%. Proficiency rates in Math: African American 21%, Hispanic 28%, white 38%, Filipino 48%, and Asian 53% (CDE 2007).

With the help of other helpers, we can design learning and support services that will enable poor, minority, high-need, and special-need students caught up in poor home, school, neighborhood, and workplace networks to learn more effectively and become contributing members of society. Hence, we have sought the help of several units within the CDE for additional information:

1. Learning Support and Partnership Division
2. Secondary, Post-Secondary, and Adult Leadership Division
3. College and Career Counseling
4. Mental Health issues: Student Support and Services
5. Substance abuse related disorders and health issues: Safe and Healthy Kids Program

This information was needed to prepare BHAMs for human development professionals in education, corrections, mental health, and health care and related fields. Hence, have also asked the government to help us obtain information regarding:

1. Vocational Rehabilitation from the California Department of Rehabilitation Services
2. Part B of the Individuals with Disabilities Education Act complaint procedures from the CDE
3. Discrimination on the basis of race, color, national origin, age, sex, and disability from the San Francisco Office of Civil Rights (June 12, 2007)

We ask for help to help us help our public schools improve the learning and support services to poor, minority, high-need, and special-need males. Hence, we lose too many of our males to drugs, violence, crime, street gangs, and so on. The data above reflects that our males are in crisis.

We met with Phillip Moore, the associate superintendent of Sacramento City Unified School District. In less than one hour, he agreed that the learning and support services we offer were needed. For example, Sutter Middle School was able to improve the delivery of learning and support services to a male student and his parent, the student's social interaction was improved, the student's school achievement was improved, and the parent's ability to meet and discuss these issues with greater levels of internal control was improved as school officials agreed to work with us to help.

Five key characteristics define the essence of a collaborative project. All are critical in building up a student's self-esteem: helping students learn how to think, helping parents parent, working with teachers, providing facilitation to school principals, and preparing community-based educators for the role to help parents and children learn how to learn and live together. People that are focused on the act to learn are better able to exert a positive sense of care that will have the most influence on the life of a parent and child support program.

This plan will strengthen the entire human system through teamwork with stakeholders: the people in home, school, neighborhood, and workplace networks to form a unified show of support.

- o Increase communication and participation in events to share information with all stakeholders in an open, problem-solving approach, focused on home, school, neighborhood, and workplace quality improvement
- o Have a strong focus on education as a top priority for training youth and parents to live, learn, think, and respond to improve their contact, interaction, and levels of cooperation with the people in home, school, neighborhood, and workplace networks.
- o Develop human system improvements by finding more ways to share resources and enhance support services
- o Prepare family members to learn how to live together and with other people and support the student's need to feel part of a learning community, which is essential to the students thoughts of success
- o Encourage parents, teachers, principals, and students to help the helper help them identify more cost effective ways to reduce

internal conflict by working as a team to improve learning for each other and with each other as a team
  o Provide incentives to parents, teachers, principals, and students to participate in programs to increase individual and group success while improving human relations

Parent-School Partnership and Accountability

For purposes of discussion on this topic, the terms *parent-school partnership and accountability* are used in this order to discuss how a parent with a child that has been hurt is handled by the school when there is a behavior issue that results in an adverse action against the child and parents due process rights. In this case, the parent notified us on July 16, 2010, that her child had been removed from class and her access to summer school was terminated.

The parent reported that she was told about the adverse action by her daughter and that the school did not contact her before, during, or after the decision was made to remove her child. The parent reported that she contacted the school and was placed in contact with the classroom teacher, who informed her that her daughter had been talking excessively in class over a three-day period, where she had been warned several times to stop.

As the family advocate, we are concerned about the parent-school partnership, which calls for the school to properly notify the parent before removing the child from school without a reasonable opportunity to respond. This was not an event where the child had acted out through signs of disrespect for the teacher, or where the child was not able to perform the required work. This was a situation of classroom management, where the teacher targeted the child on the first day of summer school. We learned this from listening to the parent and the school principal's reports during a Building Human Asset Meeting.

The objective facts are that the school did not properly notify the parent, and this act alone violates the parent school partnership. The classroom teacher did not officially write down the issues that led to the removal until after speaking with the parent, and having heard the parent advise her of intent to speak with the school principal and the district office. We learned this from the school principal's report, and that the teacher wrote an e-mail in defense of her decision to have the child removed from summer school.

The issue of accountability pertains to the No Child Left Behind Act and Title I schools that receive funds under the Elementary and Secondary Education Act for Improving the Academic Achievement of the Disadvantaged student (U.S. Department of Education 2003). The school principal is responsible for receiving the teacher's complaint, organizing the issue, and expressing the information to the parent in an understandable and uniform format, which warns the parent of the impending action, which may lead to the removal of their child from summer school. This has to be done in a simple process cycle that the parent can understand.

In this event, district policy states the parent must be informed of the options available to them. What we were most concerned about was the interaction that took place between the parent and the classroom teacher over the phone. When the parent notified the teacher that she had not been properly notified or given an opportunity to work with her to address the issue, the teacher responded by informing the parent "summer school is a privilege." The parent, feeling that the response was provocative, advised the teacher she was getting off the phone, but would follow the matter up with the school principal and the district.

On the morning of July 19, 2010, we were asked by the parent to accompany her to the school to meet with the school principal. Contrary to the view of some academics that the parents of children that display uncomfortable attitudes are not online with school policy and procedures, this case reflects that to be given an opportunity to be seen and heard makes the parent-and-school partnership a real act. The parent only had to ask the principal to review her record of navigating her children through the district to get him to acknowledge the existence of such a partnership. The school principal stated, "I have a ninth grader with some behavior issues, and I know I would expect to be notified beforehand."

Our view is that the words *no child left behind* are empty without a parent-and-school partnership that works to hold all parties accountable—especially as regard the consequences for a parent trying to work with a child that has been hurt through school. In some sense, the parent, teacher, and school principal must come together and look at the child in relation to the actual cause for alarm. In child, youth, and young adult dynamics, a summer school setting is a threatening social situation when you bring students together from multiple and different school environments. A good child or a bad child will look for strategies that will protect their standing within these dynamics at the expense of being called out by the teacher.

The classroom teacher is responsible for choosing the right child to aim a disciplinary strategy at to manage their students. In this case, the teacher chose not to review student records or make contact with the parent. Background knowledge is very important for school achievement and to learn what works best in schools that are open to parents as leaders. In this case, the classroom teacher's sense of leadership ignored the parent's need to be allowed to help lead her child. In a case like this, we learned that parents are not seen as partners. In some other cases, this may also be true, but how do you know unless there is contact between the parent, teacher, or school principal to learn from experience.

This has been an analysis of the parent, teacher, and school principal's work. Can they be considered to be a team without a focus on ways not to leave each other behind? Instead of relying on self alone, both the teacher and the school principal should have worked together to learn how to assess the child and the parent's school records. The teacher is a leader, the school principal is a leader, and the parent is a leader. This was the real problem in this case, a clash of leadership styles, but the parent knew and understood the school's process cycle.

The practice of advocacy matter most to the parents, children, teachers, and school principals that chose to work with us to improve the delivery of learning and support services to the home and to the school. Parents feel empowered, children feel the attention, teachers feel the help, and school principals feel the combination of home and school learning. The key contributor though is the advocate, in the sense that the parent, child, teacher, and school principal all need to be handled with care. Virtually all the links between home and school—which includes parenting, learning, teaching, administrating, and helping—had to be assessed and redefined as a research-based approach.

The parent, the child, the teacher, the school principal, and the helper are critically important to the success of Building Human Asset Meetings to improve learning as support services, which include home and school visits. It was because of these meetings that we were able to gather the data needed to appeal to the self-interests of each person, or group to transcend self and group interests to focus on the well-being of the child.

Chapter 2

Healing the Brain's Body

The Breath of Life

Your baby was just born. You push, and the doctor lifts your baby up out of the womb. You hear a cry. Your baby's expressed feelings activate the human system to release noise and sound equal to signs of speech. How the doctor, nurse, mother, and father responds next creates feed backward for the baby to receive. Language is used to send feelings forward and thoughts backward for the baby to receive as contact and interaction.

As you touch your child for the first time, cognitive structures form from the experience of contact. You can actually feel your baby move to interact with how it feels to be touched. Physically, you can see and feel life moving through your baby's body as each breath of air shows a sense of how it feels to have feelings. The mental aspects of new life take hold before your eyes.

You may think of these events as your baby's first acts to sense, feel, and focus their brain and body to communicate. But you have to feel these social events unfolding, if your reply is to act in response to your baby's feelings for a signs of a safe environment to learn how to deal with these emotions. How your baby feels contact sets up their sensory system backward to inform them as acts to think forward. Thus begins your baby's search for understanding contact, interaction, and signs of cooperation.

These initial life events structure how the human system—your child's brain, body, and senses—will function as a process cycle designed to perform as a green technology. All human systems have similar capacity to function in the role of a new ecology. The natural structure of brain, body, and sense events allow you to set up self, other people, and their environments for the experience of your child's goal to live, learn, think, and respond as a green technology. The child reprocesses these acts to live, learn, think, and respond.

The structures and functions of the human system and the interactive events of the man-made world are determined partly by the natural order of things and partly by the issues, problems, and concerns that affect self and other people. People cause the act to live forward to be difficult to set up for a child to experience the freedom of brain, body, and sense coordination within self. And to new human systems, it may seem uncomfortable to have contact with people that are not prepared to show high signs of care for how they experience the child's contact.

When we interact within this complexity, a newborn child may only have that instant of life to be set up for the experience of intelligent emotions or thoughtless feelings. With the breath of life, how so fast do the conditions of life take root? People may move from one point to the next without ever feeling each other or the birth of a child's new state of energy spilling out at them for signs of help. These noises, sounds, signs, and symbols of acts, events, contact, and interaction are process variables for information and exchange.

Where the essential responses to self, other people, and their environments are felt, children will find their ability to learn how to live greatly enhanced. Where such responses are not set up for the child to experience, the people in the situation become challenges to the child—sometimes insurmountable—task to sense, feel, and focus how the brain, body, and senses take on the act to live. Even the person whom is most upset and unaware of the child's need for this level of help to activate their human systems responses to their contact may be in need of this same experience—we all have to learn how to feel things.

It feels good to try, to feel things. If you want to know what it feels like to try, then receive these words. You feel the words to organize a sense of what they feel like. In doing so, you build on a capacity and potential to express how they feel to you. Now, you are set up to receive and send messages. You use language to receive, organize, and express feelings, and to try to make sense of them. Words are used to talk through feelings, and to share what it feels like to read them. We hope you enjoy our use of the word *talk* to share these thoughts forward.

What Is Language?

All learning at home, at school, in the neighborhood, and in the workplace "hinge on the use of talk to transmit words" (Slaton 2009). What this means is that the parent, teacher, helper, and child need to use these places to learn how to talk to each other's brain rather than to his or her body. If your brain, body, and senses are set up to use words that connect the act and practice of feeling things, then the words that are used to aim the things you say will be to help him or her focus within the context of how they feel.

Language is a natural object, a component of the human mind, physically represented in the brain and part of the biological endowment of the human being (Chomsky 2002). Innate to us all is the act to speak through the use of noises and sounds with the goal to make sense. This is to say, the act to speak is a building block to the way we become mental in the use of bodily acts to send feelings forward through the release of noises and sounds and to receive thoughts backward as a symbol of the central nervous systems ability to breathe through contact. In other words, the brain is the part of the human system where nerve tissue and sense organs transport signs of contact.

Recall how a baby's initial contact with the world is led by the act to breathe and cry at the same time. The act to take in air and talk is both physical and mental. Chomsky has called this "the transition from the state of mind at birth, to the initial cognitive state, to the stable state that corresponds to the native knowledge of a natural language" (2002). The aim of this chapter is to set up the connection between our beliefs that at birth, the brain is a blank organ and that all ways of learning are derived from sense experience. Such as the act to cry is a natural language used by babies to express their first senses of life: self, other people, and their environments.

In the first stages of life, language is used to show clear signs of care for how the baby has to feel. We, the people, use speech to set the infant up for contact and listen to learn if there are any irregularities in their response. Every physical response is being read as a mental and physical act to live through the events of birth. The purpose of preparing self, other people, and the environment for the birth of a child is to clear paths for the baby to receive a sense of self and other people to learn how it feels to hear noises, sounds, and words aimed at their brain. This creates ways for the baby to learn from the contact how to feel for signs of life.

Just as a baby needs a sense of feel for life, the people that support it need a higher sense of self to know how to return these feelings as evidence of thought. This is the purpose of our use of words—to speak, listen, write, and read to the brain. In this sense, language is used broadly to mean the way we set a person up for the experience of life is through the use of words to move them. For instance, in education, language is used to inform the mind. So to inform the child's brain, these responses to life help to set up language centers (Crystal 2003), or a language faculty (Chomsky 2002).

The senses play a critical role in the set up of neurolinquistic pathways between the body and brain. In this sense, nerve tissue transport feelings to language centers in the brain in response to contact. This means that the baby's response to contact is shaped by the use of language to make sense of self and other people. The central function of human systems research is to learn how to study the roles played by self, other people, and the environment through the use of talk. Hence, we must explain how the experience of talk is associated with the use of language to learn how to live with self, other people, and the environment as used by human beings.

Recall that the human body is physical. But the human system is physical and mental. To think with the human system, you must be able to sense it, feel it, and focus it. This is why we seek to learn through human systems research, which includes the environment, to study the things people do alone and together. People use talk and words to observe, listen, learn, help, and lead self and other people to feel them. In its loosest sense, language signifies the aim to ask for the recognition of feelings. You ask a person how they feel to learn about their feelings. In asking the person to speak is a demand for these feelings to make sense of their contact.

In our work with children and families in crisis, we relate any threat to home, school, neighborhood, and workplace events. These four places make up major life events where the use of language helps us to set up what we call a talking system for the parent and child to experience. Hence, the word *live* is related to life in the home, the word *learn* is related to learning inside the school, the word *think* is related to thinking inside the neighborhood, and the word *respond* is related to responding inside the workplace. This makes up what we call in human systems research, an environmental learning system.

In this sense, we are setting up the experience of talk. We observe how the parent uses talk to make contact with their child to learn how it causes them to feel things. As we listen to the parent and child's use of talk, we are writing down the words and their connections to the feelings that are being displayed, as send and receive paths between them. We have learned, from their use of words, how they may feel, when it seems as though one or the other is not being seen, heard, or felt. There may be anger in their eyes (emotion), there may be signs of fear in their movements (feelings), or there may be signs of anxiety (tension).

The words we use to aim our talk are intended to help a parent and child feel for a sense of calmness. Can you feel how learning to live with each other through crisis may be helpful? The parent has to live through the crisis of parenting a child that is growing up hurt and the child must learn how to learn to receive words that may lead them to change. This is the problem of language. The words you use can either open the person up to your contact, or lead the person to resist their need to interact with the words to grasp a sense of feel for what they mean. A child may not have enough experience to learn how to live with other people without a grasp of how it feels to have feelings and thoughts from the experience of participating.

Reflect on how you use language to observe, listen, learn, help, and lead yourself to feel for signs of understanding acts to live, learn, think, and respond to contact. Each word means to feel for signs of care. When you lack the experience to care, you are more likely to block the use of talk. This means you do not want to feel the experience of the word. Each phase requires the act to send feelings forward and to receive thoughts backward. This means that the words your child aims at you are loaded with feelings they need responses to if they are to learn how to think through the experience of talking. What a parent says to a child are acts to send their feelings forward as well, but if they are related to the child's talk, then they are thoughts to the child's sense of feel. The parent's feelings have to be responded to in similar ways for thought, as they begin to penetrate the mind's eye.

When language is used in this broad sense, it is usually centered to cause the technique of speaking to take place. When you express the word *live* or the word *learn,* they mostly relate to acts in the home and acts in the school. These words are used to focus the parent-and-child relationship on learning how to live and learn alone and together from where they are in the crisis. In this sense, they are intended to organize signs of care. To move a parent and child out of crisis, they must be set up to live as a family and learn how to use education. These words lead to feelings and thoughts about the experience of home and school.

They may produce negative feelings and negative thoughts, unless by chance—the people there—are using talk to lead each other to live and learn together. Such feelings and thoughts about family and education are an efflorescence of sense experiences. For this to happen, the parent and child's feelings for and about the experience of school have to be improved to set the stage for thoughts to come to mind, signs of controlled emotion in the experience of talk aimed at their brains. The people at the school would have to show high signs of care for the experience of their contact. This means that words used to talk to the parent's brain or the child's brain would be aimed respectively in relation to their roles to observe, listen, learn, help, and lead.

This is the power of language. In sense experience, how words feel, set up or breakdown, send and receive paths between the body and the brain. In the next act to send feelings forward, the brain and body may either work as a unit of action or fail to work as a human system. This means every word we use plays a role in the event to aim the brain to accept the body's need to act through these feelings to live. To live with these feelings is a natural human learning goal. This mode of learning receives special attention throughout this book.

This is why when we say the act to live, we are referring to Progressive Investing. Progressive Investing is a technique used with human systems research to set people up for acts to live, learn, think, and respond to become more informed. We want the parent and child to feel for their need to become more informed to change routine patterns of behavior and emotion, through human learning events. This leads them to experience a sense of self, other people and awareness.

Can you feel the words *progressive investing*? To grow, mature, and develop means to advance from one point to another. If this is the meaning behind the words, Progressive Investing is the act to move through. It is the practice in the act to learn. The word is meant to help the parent and child set up his or her aim to learn from what is done to move through the act.

Progressive Investing is a branch of green science that inspires the act to live and learn from the experience of awareness. To encourage parents and children to live and learn through hurt feelings, they are asked to think about these words: *grow, mature*, and *develop*. How do you feel? Each act to live each day is to grow the experience of how it feels to have these feelings. Each step is measured in terms of how these feelings are being dealt with through their contact and interaction with us.

We are the models the parent and child learns from; our acts to live each day to learn how to help them help us. Learning how to live is a central term in Progressive Investing, because it covers the home, school, neighborhood, and workplace networks being used to aim talk, reading, writing, speaking, listening, and helping skills. This means the word help relates to self, other people, and the environment bounded by speech to identify and to recognize the need for goals. For instance, to think is relayed as an environmental act since to do so is a sign of feedback.

Perceptions of this kind are feelings of thought. You can feel me trying to talk to your brain. Recall how a baby cries outward and that we the people, respond to send feelings inward to change this effect. In fact, we are trying to change the baby's feelings to inspire signs of thought. The formal process of Progressive Investing leads the act to live each day to learn how to respond to the baby, self, other people, and the environment. Words are used to aim all contact, for speech and clues to identify the event to live and the goal to learn from the experience.

You can feel for feelings of thought. Hence, a baby cries to feel for contact. It is the experience of contact that causes the baby to grow more aware. You can feel the baby's contact as feelings of thought emerge from the interaction. Feelings are being processed; the parent and baby are more aware of each other's meanings through the use of talk and words. This is the basic goal. To think through the way talk and words feel.

We live in search of safe places to learn how to feel things. In this sense, humans are green learning systems set up to reduce the waste of land use. Our brain, body, and senses help us to grow more aware of the words that lead us to feel safe. We experience these words to learn how to move through the practice of bringing thought to mind. Language brings thought to mind through the experience of feelings: where to live, where to learn, where to think, and where to respond as the brain quickly grows aware through words that lead us to feel for safe places.

Language is the sound of noise being transformed through contact and interaction to produce feelings of thought as signs of meaning or not. "Two crucial factors which interact with and reinforce all of the activities taking place in the home environment and which have been shown to enhance the child's intellectual development is the quality of the mother's linguistic interaction with the child and/or her encouragement of the child's linguistic advancement" (Wilson 1992, 66). This means the use of words that lead the child to feel safe at home, at school, in the neighborhood, and in relation to the workplace is crucial to their future linguistic interactions that will inspire signs of care.

What is Cognition?

You learn from the practice of Progressive Investing how to test the experience of action.

Each time the learner chooses to learn with us, they are choosing to act through the experience of our contact. This is because we are part of the experience, the school of thought. Each time the learner chooses to take action to learn with us, he or she is doing this work to become more informed through human, cognitive, and behavior science (Slaton 2009). This is to say, to take action to live and to learn is a physical and mental response (p. 36). The experience of which is cognitive.

Language is the use of words and speech to send feelings of contact forward from brain, body, and sense events; and cognition is what takes places as these acts are being experienced in a backward feed cycle. A child's sense of knowing is connected to how they come to think through responses from the environment. This means that a cognitive (building) block is created each time the child sends feelings forward from self, or receives thoughts backward from other people or the environment. A block is a structure that forms ways of knowing through proof of the experience.

A child may cry, but stop each time the parent uses talk to find out why. The experience of the parent's talk is being felt by the child as feedback, which produces a sense of awareness. It can then be said that each time the parent uses talk, the child looks for certain feelings from the parent. Cognition is connected with all stages of life and is thus a much broader event than consciousness (Capra 2002). According to Capra, awareness comes from lived experience—built on certain stages of cognitive complexity that use the brain, body, and senses and a higher nervous system.

The brain, body, and senses form interdependence through the use of language and cognition as experiences. The way they belong together in a pattern of contact is made more intimate, and yet more reflective, than any separate acts of feeling and thinking could be effected by internal and external events. When we talk about the patterns in one's feelings, or about patterns of thought or patterns of experience, we are referring to the internal and external process, and what we know about brain, body, and sense events being flexible, not fixed frameworks. This is why the act to live and the act to learn can be seen as a unified pattern of response to contact.

You can see the child responding to a call for them to come to breakfast with their hands washed as they may have previously been informed. You can look on the child's face and feel how they may be feeling about the act to live through these feelings. You can read the child's facial expressions and sense how these experiences are interacting with their brain and body. You can also ask the child if they got enough sleep last night to learn about the thought process of these feelings. These are all patterns of dynamic states of awareness, the actualization of which depends upon a buildup of experience along a continuum.

Recall that Progressive Investing is the start point in the act to experience the continuum. You live each day to learn how to move through the act to participate, experience the event. Hence, the child builds on a capacity to move in relation to how they feel things. To feel something is an event to the child. To live through the way things feel is an experience of an event. To learn how to move through it is Progressive Investing for an awareness of it. This is called learning green, to advance the idea of the child being an environment in the act to learn how to live with a sense of time.

Capra (2002) describes two types: Primary awareness arises when cognitive processes are accompanied by basic perceptual, sensor, and emotional experience; and higher order awareness, which involves self-awareness through a concept of self, is held by a thinking and reflective participant. In other words, the child has to be open to the contact, which means the elements of language and cognition need to be neutral to each other and have to be held together through the experience of external interaction. Hence, an organic transfer of thought, in this case, is a sign of synthesis from the process of feelings received through external contact.

The child feels the exchange, because the experience of feeling recognized produces it. One of the more important aspects of an internal to the external, to the internal effect is the fusion of experiences. Interactive processes are able to produce higher process cycles and therefore are able to produce new blocks and structures from the experience of experience, of the experience, and hence, these processes may be observed (Taba 1932). When the child interacts with the parent, the act to sense, feel, and focus, the contact is observable as acts to live. When the child interacts with the parent's response, the act to receive, process, and respond to the contact is observable as acts to learn. The experience of the experience is the experience of feelings, thoughts, and awareness.

This means that each state of awareness is created from a special qualitative feel to deal with the dynamics of becoming. A full grasp of the human system will be reached only when we approach it through the interplay of three different levels of description—human systems research of observable events, the laws of physics and neurobiology, and the dynamics of complex systems (Capra 2002). Hence, the ordinary forms of growth, maturity, and development we refer to in the process cycle of life, space, and time can be defined as patterns of becoming a more structured and functional human system.

In human systems research, we learn that the child has to be set up for contact and that the parent has to be set up for interaction. These two acts produce signs of energy that reorganizes the release of new energy. In other words, the child is more open to feeling the contact and the parent is more open to thinking through the child's interaction, which reorganizes not only how they release new energy, but also how they act to live and act to learn from the experience, which produces changes in behavior. Certain feelings relate to structures in the brain or connect to memories, and certain thoughts connect with external events or to people activating a desire to work through the complexity.

The child and parent may have come to us for help three hours ago, but because of the way the process of learning how to use talk and words was working for them, they continued to work far beyond our expectations. The child would sit quietly as the parent spoke, and the parent would sit quietly as the child responded to the words they chose for them to experience. As we continued to observe, we realized the difference is essentially in the release of words, which activate capacities that were not previously structured for these functions. They were physically controlling their bodies, as their words were being exchanged and interpreted by us as signs of growth, maturity, and development. Hence, they were using talk and words to move through their pain or bad feelings, long enough to experience new energy.

Learning green is viewed through our grasp of the parent and child's act to learn with us, which reflects a sense of time and space. The parent and child will move to feel for each other through Progressive Investing! The term Progressive Investing relates to the use of energy to take action and to feel for ways to move through it. How do you move from the unknown to the known? The parent and child looked for signs of meaning in the experience of contact. The parent and the child aimed the words they used to talk through, at each other's brain. Hence, less time was wasted.

These new noises and sounds make up words that enter the cognitive process as new ways to communicate, and being, new forms of talk will transmit changes in the flow of feelings between the brain, senses, and body. The process of new energy coordinates the experience of experience of care as a continuous and natural process of brain, body, and sense coordinates. Cognition is thus the release of new energy which activates the capacity and potential that was up to that time, blocked by social signs of hurt. When a child cannot look you in the eyes, when a parent cannot stop talking over the same issue, these are social signs of being hurt.

The reason why; the way the child and parent uses language changes the structure of cognition is because the flow of information is sent out through feeling and thought processes and the capacities producing the first are more open, hence producing more potential, than the capacities moving into stages of the new experience of experience. The experience identifies the meaning of becoming aware through the structures and processes which are elements of cognitive experiences. The child is now sitting still, body aimed in the direction of the speaker, and the head and neck is following. The parent is talking to the child's brain, and not to the child's body.

These are the aspects of language and cognition we found in each of these cases, something new was produced in the expression of human feelings and thoughts. The parent and child chose to listen and observe the sounds to learn from the words, how to help each other lead the other through the use of talk. In each, the process is that of learning through the experience of feelings and thoughts. This is a process of the act to live with each other from where they are, in the crisis that functions to form a more fused, complex feel for each other's act to think. These are social acts, we learn from.

Progressive Investing is the experience of the affect, which we work to structure across our continuum. The parent and child are moved by the contact to interact, which means to store memories of it. As the first step is a search for cause, and the second step is a search for relativity through acts to feel for signs of the effect. Cognition is the effect of experiences being felt in a field of experiences. The science of Progressive Investing takes form over time.

The following Progressive Investing Model frames how the practice relates to the parent or child's feelings, which will reflect progressive or regressive acts in their real lives within a set of boundaries.

Progressive Investing Mode 2

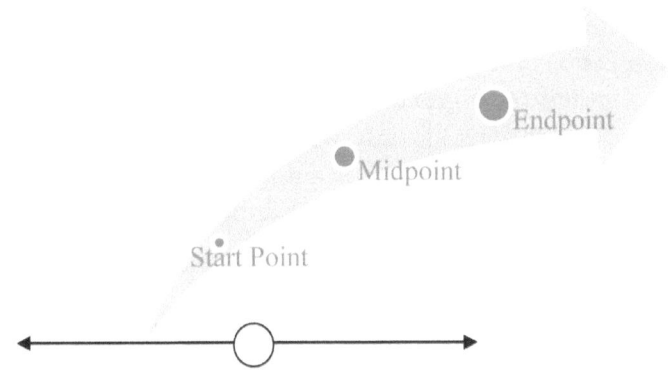

How the child and parent grow is formed by the changes that occur. How the parent and child mature is shaped by the signs of feelings and thoughts being shared. How the parent and child develop is influenced by their act to live each day to learn how to become more informed about how they feel and care to think. These stages are all determined along an action learning continuum, where the past identifies the start point, as the event. The process of ways to learn is the goal or endpoint. Hence, anything between these two points describes the affect; we choose to call it the midpoint.

Learning how to become more aware of safe ways to use the land, is green science generated from the parent and child's act to clean up their space. Space that may be occupied by self, other people, and things when focused upon in this way, is learning green. Learning how to live each day to learn ways to become more informed through the aim to improve the things you can are physical acts. Hence, learning how to show signs of care is physical.

What is Physical?

"In human systems research, human systems physics is the study of the self in relation to being physical: in the mental, the social, the emotional, and the environmental construction of the whole" (Slaton 2009, 72). Each event connects the relationship between the act to live and the process that includes how we learn from experience. Hence, the goal is always the act to live, which is physical (p. 38). This is why the study of how the parent and child learn is a physical science.

"Sometimes when teachers discipline students, the teachers can become very frustrated and emotional. These situations can cause them to overreact by grabbing a student or being physical in other ways" (Boynton and Boynton 2005). You can feel from these words, the physical acts of the teacher looking for signs of control, or you can develop thoughts through your observations of these words that the teacher was thinking without feelings for the student. The act to discipline the student or the act to get the student to show signs of discipline are both physical acts. Either of these acts will stir up feelings.

"The term physics is derived from the Greek word and meant therefore, originally, the endeavor of seeing the essential nature of all things" (Capra 1999). All things that can be seen through the eyes are physical. The human being is a physical thing with a body that feels things. Hence, you can sense from the statement above that the teacher was emotional at the site of the student not doing something to their liking. You can feel the teacher feeling the student, but you cannot feel any signs of focused thought. Hence, the teacher was not feeling the student. The teacher was thinking without feelings for the act of discipline being requested from the student.

You can believe then, that the teacher did not use a sense, feel, and focus cycle to interact with the physical signs of resistance being displayed by the student. Being physical, and what is physical in this sense, is determined by how you react to the things you can see. When the body makes contact with something through the senses, you can feel it. This means there are feelings in everything we experience, a sense and feel cycle. When the teacher used thought before feeling for the student's contact, there were no signs of focus in the reaction. Hence, the teacher loss control of the brain.

This reveals to us that signs of discipline are in the physical contact under which the interaction between sensing and feeling things take place. If the teacher focuses the way the student's contact feels, then this is a sense of discipline. Meaning, the teacher's actions were more likely to flow from sensing, feeling, and focusing how the contact feels as the experience to experience thought unfolds through the student physics. Our perception follows a symbolic relation rather than a casual one. We sense that the student is acting out, or why the teacher would be angry. The signs of these feelings are fragments of potential reality that point to other conditions as equal signs of other things. Other people were most likely watching these events, and the teachers concerns were for their meanings and connections to them; the other people, or the events.

This wording of the situation does not discriminate between the physical and the physics, between the student, teacher, or other people. Hence, we can learn from the experience of contact that has to move into stages of interaction, the value of the sense, feel, and focus process cycle that has to be followed and tested each time. The field of physics is the field of experiential events. For instance, we can draw from these words, because they are being presented in the physical sense. Hence, these words make up an event. You can feel them and you can focus how they move toward states of awareness through the experience of trying to make sense of them. To feel an event is to focus the goal to make sense of some affect, which functions in the flow of interactions.

Hence, the teacher has to realize the true potentials of being physical and working with other physical human beings. The body is a symbolic thing to feel as an event of the senses to focus all signs of care. To think of the body as an event, with a goal to have an affect which is designed to influence the feelings of any observer has to be acknowledged. Acts to sense, feel, and focus the brain and body is fused in the subsequent flow of new events. "Any element of perception functions as a sign for all possible different meanings, and those meanings depend on the response patterns that are called out by that symbol" (Taba 1932).

Physical action learning is a perceptual term used to visualize feelings through the use of words. For instance, to store information on how it feels to send feelings forward through the use of print: help me. You can visualize me as the helper, and your response to help me receive feedback from these printed words. You can learn how to live with me through the use of these words to help. How you help is stored by the affect that is in making contact with the words, and how you interact with them to visualize how you feel.

Words are used to send a message that will help you draw symbols that reflect upon their possible meanings. If you are using your sense, feel, and focus process cycle, then you can imagine the whole event as a performance of thought in relation to feeling things. You now know how to feel for signs of new energy. The physical act to learn through the experience of feelings forms relative to thought. At this point, how you feel is how you think. The only difficulty that remains is in your act to believe what is clearly observable in your sense of mind.

In the experience of awareness, only those who can feel for it, come to focus on it, and to learn from it, there must be thought reached through the experience of contact and interaction. These are patterns of sensing, feeling, and focusing thought. Hence, for this reason, we can think about how the brain feels to have thoughts moving through the body with the physical signs of discipline as an object of this focus. How the brain feels things for an influence is a material act, for the body is physical, which means it can be seen in a state of focus, for all who have eyes to see.

What has been said about physics and physical is that they represent symbols and symbolism as true stimulus for emotions in general. The interaction between sensing and feeling a thing that the body is in contact with is an emotional experience that cannot be hidden from being seen when it's public. Each time the body moves, each time the mouth speaks, each time the brain is exposed to these feelings, sense has to be used in a way that coordinates these natural signs of life. But for the purpose of discipline, that is physical control of the body through the use of the brain to feel the contact and interaction, focus has to become the general experience of this affect.

The body does not control the brain, because the brain controls the body (Amen 2010). "Fifty percent of the brain is dedicated to vision" (p. 3). The brain may control the body, but raw feelings can affect vision. Hence, the brain controls behavior through the practice of thought. The senses move feelings through a physical process to control them. To sense, feel, and focus are physical acts of the body.

A disability then, is when there are sufficient signs in a pattern of interaction that lacks signs of thought. In this sense, a learning disability, or a learning problem is studied in relation to a student's ability to sense, feel, and focus the brain and body to receive experience. The self is considered in relation to contact and other people are considered in relation to interaction. The environment is considered in relation to the influence of home, school, neighborhood, and workplace networks as symbolic places.

"They belong to the type of experience where events and their qualities are classified and systematized" (Taba 1932). If a student cannot show signs of learning from the experience of interaction, then there is a problem in the sense, feel, and focus path. These paths form intellectual experiences through their dynamic restructuring of feelings in thought. Hence, intellect is a result of successful cognitive processing. When a student can send feelings forward, but not receive thought backward, this means they are not recognizing the teacher's physics.

The student has to recognize the teacher in the physical sense, to feel and focus the contact toward paths of interaction. In other words, if the student cannot experience the teacher, then the path of focus is broken and there are no sufficient signs of thought in the reaction. Being unable to respond to the teacher's instruction is associated with not being able to send or receive physical or symbolic images. For instance, the student's physical state is disturbing to the teacher. The teacher's perceived physics are not being taken in by the student.

"Being able to do what others do in a classroom, in the way the teacher wants you to do it, is programming" (West 1997). The student cannot feel what they cannot or are unable to experience. In this case, it is the programming of classroom interaction or rules of engagement in general. If the student could sense, feel, and focus the experience of classroom discipline, then they would. But if the teacher cannot inform the student, then the student will be a sign of frustration for the teacher to experience. This is a form of disability or of problems learning from physical contact and interaction.

These concepts—sense, feel, and focus—are related to human experiences in home, school, neighborhood, and workplace networks. The experience of being physical as an event with these qualities is classified under human systems research to learn from self-action, other people's action, and their environmental interaction. This way, the types of experience they are experiencing can be turned into a formula and related to acts of becoming physical. Hence, we can read into the words used to describe the student and teacher's behavior as acts of physics. As you have read, the experience carries the solution toward the events they represent.

The idea of how to change a learning problem is mental. The physical act not to sense and feel the contact controls the behavior pattern that needs to be changed. This is why the path of focus has to be connected to the physical act to live though the contact, which proves the brains involvement in the process to be mental. This is what the process cycle implies—knowing how it feels to be visible; knowing how it feels to see; and knowing how it feels to aim the brain connects sense and receive paths to sustain levels of control over the body.

What is Mental?

The parent and child learn how to take action to live each day to become more informed through the changes in their behavior toward self, other people, and the environment. Moving from the physical act to the mental processing of the act is mentalism. "For this purpose, mentalism starts from the point of connection between space, energy, and time studied for awareness: where am I, what am I doing, and how long have I been here in this state?" (Slaton 2009, 78). Any study of what it means to be mental that does not seek to relate physical events as products of the senses, which connect the parent or child's needs to the mental aspects in a response to explain how events and processes change right and wrong ways to react, strip away cognitive meanings.

Mental is the process cycle itself, as to become more informed. "Physicists have come to see that all their theories of natural phenomena, including the laws they describe, are creations of the human mind" (Capra 1999). In other words, mental states and processes exist as separate events that create and change behavior, and are used to explain human nature. The brain and mind is not the same thing. "The brain is a physical organ composed of atoms and molecules that inhabit the skull. The mind is much more than that; it transcends the head and operates throughout the whole body" (Sousa 2001). The following Human Systems Model reflects on the process.

Human Systems Model 1

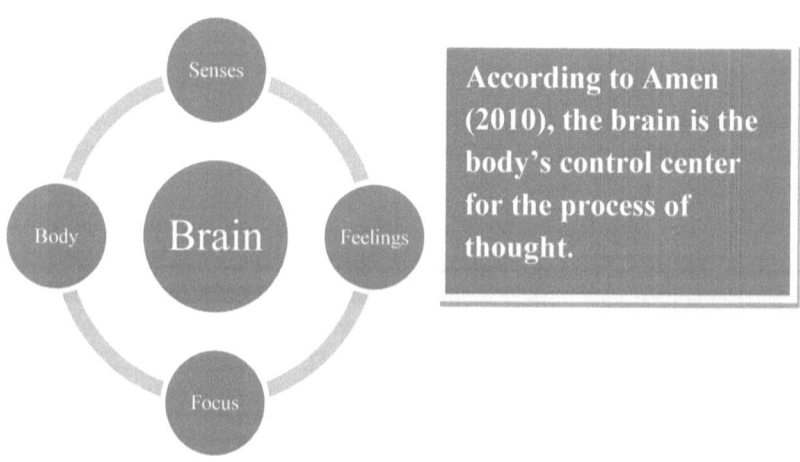

We all have a brain. But do we know how the brain works to become mental? To understand this work, we have used such words as sense, feel, focus, and body. We want to explain how the brain becomes mental through the use of these words as a process cycle. Everything that happens to the body is transferred to the brain through this process. You talk to me, my ears feel for the meaning in your contact through a line of focus. The body is the physical structure that hosts the brain, senses, feelings, and focus capacities and potential.

To sense means to feel any bodily contact through nerves that connects it to the central nervous system or the brain. To feel something means to recognize a sense of contact with the body for transfer to the brain. To focus means to center how the body feels the contact through the senses for transport to the brain. The brain is the central nervous system of the body (Audi 2001). To become mental, a child has to take action to live with a purpose in mind. To do this, the child has to know they have a brain to use in the act to live.

To become mental, the child has to learn from the act to live, ways to use the senses to learn from the experience of contact. In this case, learning involves the act to live through the experience of contact. The contact between self, other people and their environments are experiences of events that lead to these acts to grow and become more mental. If the child is aware of this need to grow and become more mental as they experience language, cognition, and being physical then the child knows their act to live connects the brain to their body. They can sense it, they can feel it, they can focus it, and they can live each day to learn how to become more informed about how they grow to become more mental.

As the child's brain is growing and becoming more mental they show a higher sense of awareness for academics; reflection; experience; needs; purpose; analysis; behavior and so on. This approach is based on the child's innate need to learn how to live in a home, learn in a school, think in a neighborhood, and respond in a workplace. The child's brain has to be connected in these places to become mentally aware of them. These acts—to live, to learn, to think, and to respond—structure these events, learning in home, school, neighborhood, and workplace networks.

The child's senses of reality grow from their experiences of what they feel they can believe, desire, process, aim, hope, or focus for. The Human Systems Model shows that this is a practice.

Human Systems Model 2

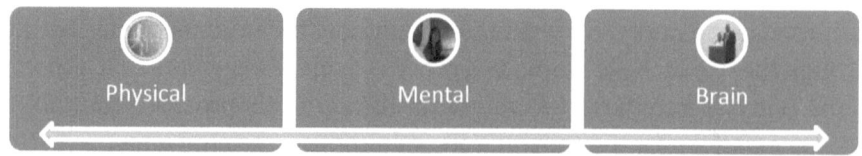

The body is physical. The sense, feel, and focus cycle is mental. The brain is connected to the process cycle through neurological and mental events. We know this, because we have been working with Donny since he was 6. Today Donny is 12 years old and has moved from kindergarten to the seventh grade with us as his helpers. We have observed Donny grow and become more physical and mental through his act to live each day to become more informed about being physical, mental, using his brain.

Every physical event is neurological. The sense, feel, and focus cycle centers the contact for transmission to the brain as a mental process. However, the child has to be aware of these hidden connections to their brains capacity and potential to grow and become more mental. This means that the brain is in the body to help the child learn how to become more mental through their experiences with contact and interaction that causes thought. "In its next sense, thought denotes belief resting upon the basis, that is, real or supposed knowledge going beyond what is directly present" (Dewey 1991).

"All thinking arises out of concreteness, which means out of the brain patterns resulting from actual body movements of interacting with actual things. But thinking then moves toward autonomy, that is, moves toward independence of those concrete patterns or physical principles" Wilson 1992). Thinking is the mental process of growing and becoming more of a human system. This means the child has to understand what it means to be physical and how it feels to have a body that is mental. This is what teaching the child how to act through the experiences of living, learning, thinking, and responding to self, other people, and their environment implies.

The child has to learn how to come to think with a real sense of self, other people, and the environment in mind. As they grow, mature, and develop with this knowledge moving them they enter more stages of learning how to live inside their bodies, with a higher sense of feel for self, other people and the environment to focus how they learn to change their mental states of mind. In other words, the child learns how to use self, other people, and the environment as physical events to experience. In the event of experiencing these feelings the body becomes more mental in the process of how they live in a home, learn in a school, think in a neighborhood, and respond in a workplace network.

The neurological basis of the central nervous system is to feed sense data forward to the brain through its experiential connection to feelings that need to be focused to react in the formation of thought structures. The child has to learn from the initial contact how to move through it. Moving through it means to interact with the contact as feelings are forming connections with the brain, and the need to think before the link is completed creates tension. Hence, the initial processes of becoming more mental than physical can be crossed up by a lack of focus.

This is the Progressive Investing approach to preparing children and parents for contact and interaction with the process of learning how to live alone and together. We have deposited a set of theories on the use of language, cognition, physics, and becoming more mental, through sensing, feeling, and focusing body and brain events. This is the nature of becoming a more thoughtful human being. This makes the body and brain function more like a human system. In this case, a more structured and functional human being. Hence, the child should learn how to study life inside the home, to better prepare for learning outside the home.

The body is physical and houses the brain. The brain is mental and lives inside the body. To prepare the child to live in the home, in this sense, is the same thing as to prepare the child for life inside and outside their own mental body. Built into this theory of action learning the child's true nature is being carefully moved through the use of special words and events—this is to help them learn how to observe, listen, learn, help, and lead self to look at the inner and outer processes, that take place as they are learning how to live with us. Becoming more mental than physical is a hidden process cycle they must come to experience in space, energy, and time.

The child's body is physical, but the brain and these process variables are hidden beneath their skin. We need to help every child learn how to become more informed about these connections to becoming a fully structured and functional human system, which is in their very nature. Hence, we do not believe the child's brain, body, and senses are designed to make them function as machines.

This is why the brain's body is wireless. The child is supposed to live to learn how to use the human system to do things. The mental states of a child grow with the freedom that is allowed. Freedom in the sense of becoming more than what can be seen or imagined. But to feel things from the inside to the outside connects the mental process. Hence, the child experiences things to learn from the experience of self as a technology. The child's brain, body, and senses are designed to set them up to become information processors. As such, a child can learn to change how they receive sense data, process feelings, and respond to focus as information moves through these connections.

What they say and do next, is called creativity. This is what the word becoming implies. This is a state of change, the child changes. Becoming mental is a process of changing from the inside to the outside of the body through acts to live, which lead to learning, thinking, and responding to the initial experience of this purpose. The child's changes naturally start as being physical, but the fact is, they are mental as well. Becoming physical has to be neutralized through the process of becoming mental. One event represents signs of discipline, and the other represents signs of self-control. The child's body changes with signs of discipline, because discipline is physical. The child's brain changes with signs of self-control, because self-control is mental.

Last but not least, you might imagine that the child is unaware of a great many of these physical and mental events; in this case, growing and maturing is taking place where the eyes and sense perception is used to feel these events. For example, why is the body physical? Now, how do you learn why? You might not expect the child to use a system or strategy to look at the body as a mental process, but then, this is what their human nature causes them to do. Their eyes, ears, mouth, skin, and nose is used to feel for a sense of what the body is. We will see in other chapters that a complex level of sensory organization of these events take place, as a higher mental process of awareness.

For these and other reasons, the human system is structured to function as a technology from the child's capacity to grow and potential to mature as a learner. In fact, human systems research moves us to believe that these cycles of growing and maturing as a technology explains how the child may learn to deal with the complexity of becoming mental in home, school, neighborhood, and workplace networks. For instance, as the child becomes a student in the study of self, other people and the environment they grow, mature and develop with their brain, body and senses.

What is Social?

One of the most important aspects of being mental is captured by the act to sense and feel contact, in the process to focus ways to respond. With action, being physical is objective. Moreover, being mental is not the whole act; and because of this, being social creates a total effect on the act to focus (Slaton 2009). The act to focus the brain, body, and senses reach across physical, mental, emotional, and environmental domains of human systems research, which connect the parent and child's social need to respond to how they feel.

Social is seen as physical, where two or more people enter contact which may lead to stages of interaction in the environment, or between self and other people as forms of contact and interaction reach signs of participation and/or skills through sense events. Social then, is the physical act of being seen, felt or experienced in the same sense of interaction. Physical action involves the contact between self and other people in the social process of interaction as mental events. In other words, the parent and child can learn how to live alone and together, because they are physical; their contact can be learned from through the way they interact with each other as social agents.

Hence, the nature of being social is a mental process of physical contact and interaction. The parent and child's bodies are physical, meaning they can be seen, felt and experienced in the brain. A parent learns how mental their child is from the way they enter and exit social contact in relation to an event or goal. The event is to become physically aware of the goal, which is to learn how to interact with self, other people, and the environment. To do this, the child has to take action in relation to the contact and be seen doing something (reading, writing or drawing).

To move through social contact with signs of interaction, the child has to cooperate by experiencing the event (read, write or draw). The child's action can be viewed in relation to the parent's goal, which is to learn how mentally involved they are in the event. Since, the child's body is physical; it is a social object that can be learned from. However, the parent can go even farther into forming an opinion of how their child responds to tension, stress, or pressure, by asking some questions (Slaton 2009). Now the communication signals are more obvious and the levels of participation are determined by how the child cooperates.

The child is learning how to be social, and how to communicate socially. In this sense, the child is asked how the contact feels, and in the mental process of feeling the interaction to the contact they show signs of a will to participate in these social events. With respect to learning how to be social, the child and parent must learn together to reach higher stages of cooperation as the child has to choose to participate in each event. Each event the child participates in, the child is being asked to display the effects of the parent's goal. To this end, the child and the parent are practicing the social development of participation skills.

Participation skills form from a sense of self (personal), other people (academic), and the environment (social). The personal self wants to understand the people's use of talk in social places. Having a sense of the people means the child is in the act of learning how to work through contact and interaction using language skills. This in turn means that their sense of self is becoming mental through social events in their home, school and neighborhood networks.

"As powerful as the social self-image is, it alone does not indicate the likelihood for future success in a high-tech society. Nor does it determine the extent to which youth are likely to be motivated to achieve success in mainstream America. It is the academic self that dictates how well children will fare in a society where survival will require high-order thinking skills and other academic competencies" (Kykendall 1992). "Unfortunately for Black teenagers, those cultural stereotypes do not usually include academic achievement. Academic success is more often associated with being White" (Tatum 1997). A parent has to focus their goals to include social events, set up to show their youth how to develop these participation skills.

This is why your youth has to learn how to participate with self. Having a sense of self in relation to other people and the environment helps the youth learn how to help oneself through these physical, mental, and social events. Your youth cares about how they look, and feel how people view them in day to day home, school, and neighborhood situations. This is how your youth learns to compete with a high sense of self in social places. Your youth is learning what it means to be physical and mental in social events that require the act and skill to participate.

Your youth has to be able to use talk to communicate by sending feelings forward with the aim to receive feedback from the contact and interaction that occurs to score the levels of participation they create within self and in relation to other people to sense and feel the affect as they become more focused. This helps to explain the socialization process in families—for example, socialization in terms of symbolic validation of self. The process can occur through positive defining in which significant others articulate a positive definition of the self, such as how bright your youth is" (Logan, Freeman, and McRoy 1990).

"The way people respond to a family (group) problem—how they see themselves within it, how they organize (or do not organize) to propose and achieve solutions—is part of the problem itself. In other words, the nature of group process-specifically, the need to establish democratic group processes as the means to achieving participatory democracy—needs to be addressed" (Appelbaum 2002). The parent has to learn how to teach participation skills to their youth in relation to home and school. Parents may learn how to do this work through our laboratory schools that will be presented in later chapters.

However, as you can sense from the order of these words—*language, cognition, physical, mental,* and *social* in relation to contact, interaction, and cooperation as being qualifiers of participation; we think these skills are comprehensive and hence, complex. What we are saying is that your youth's capacity and potential to choose how they use the senses to feel and focus contact and interaction has to be structured to function within a system of strategies; all of which they can reject, or are resisting. By setting up a system to help your youth learn how to live with you, will help them learn to make these choices with your image in mind.

Progressive Investing is a social act, in that, the parent and youth are asked to take action in relation to contact and interaction with other people and their environments to learn how to live each day to become more informed. The Progressive Investing Model below expands on the practice to become more informed through the structure of learned events.

Human Systems Model 3

Because being social is such a physical and complex act, with many effects, we have chosen to organize our system and strategy around four basic concepts in self-awareness. Too add a few more key words, many youths are in need of a higher sense of self, which is a personal, academic, social, and occupational synthesis of four fields of experience (Slaton 2009): home, school, neighborhood, and workplace learning. Learning how to live in a home is personal; learning how to learn in a school is academic; learning how to think in a neighborhood is social, and learning how to respond in a workplace is occupational. Each experience of the event is social in the act to become more mental and in the goal to become more self-aware.

In human systems research, you reach for a time of: an awareness of self and reality and their interaction as a positive value in itself should be present in the research processes (Bentz and Shapiro 1998). The practice of Progressive Investing allow you and your youth to learn how to become more aware as you live each day to study and make sense of who you are, why you are, and what you are becoming in the process. We believe that as your social knowledge of this work expands the capacity and potential to assess self, and study how you live will more and more come to be seen as participation skills or a system of strategies, that help to organize a higher sense of self and world.

This means our Progressive Investing approach will allow the reader to assess and gauge the influences that affect a parent and youth in crisis or that has been hurt by major life events social behavior. For instance, some school principals, teachers, employers want to move away from labor (parents and children) that appears to be too upset to sense and receive information and knowledge due to being hurt or caught up in a crisis of negative feelings. Hence, the parent and youth with full blown signs of being in crisis or hurt are not expected to be able to learn how to participate in society.

However, when you consider that our Progressive Investing approach prepares the parent and youth to move in relation to contact and interaction to learn how to cooperate, these are social signs of an effort to participate. For example, when the parent and youth meet us at their school site to discuss school related issues they are in the act to learn. When the parent arrives, this is a social act to participate, and when the youth enters the meeting with a calm state this is a higher sense of self at work.

The following Progressive Investing Model shows that the home, school, neighborhood, and workplace network are fixed social symbols we set up for the parent and youth to experience through acts to practice.

Human Systems Model 4

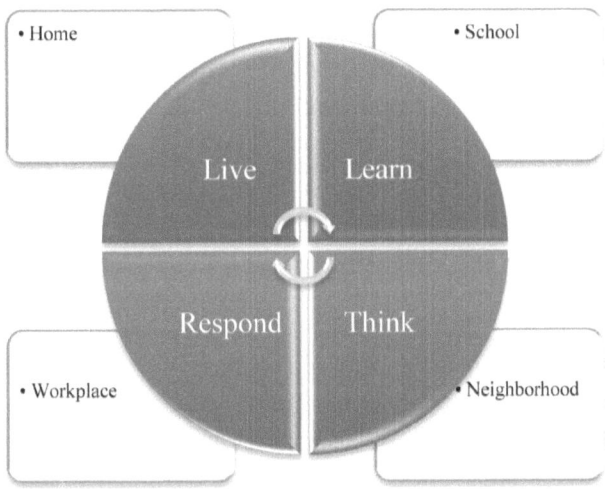

Learning how to live in a home is a social act. Learning how to learn in a school is a social act, learning how to think in a neighborhood is a social act. Learning how to respond in a workplace is a social act. When you and your youth take action it is to experience the field of experience to live with, learn with, think with, and respond with each other's image in mind. The image of a higher sense of self and other people is what helps you and your youth sense, feel, and focus these acts in the process of personal, social, academic, and occupational skills.

This is a social learning approach to gaining the experience you and your youth will need to cooperate using this system and strategy in the home and cooperative skills in the school, neighborhood, and workplace network alone and together (Johnson and Johnson 1994). Being able to cooperate is more than a central theme. In this case, it is the decisive variable in learning to successfully participate with self and other people. If you or your youth cannot sustain higher physical and mental processes of cooperation then your capacity and potential to participate is reduced.

Indeed, the potential to participate in social events within the home, school, neighborhood, and workplace network can be upset by anyone who lacks a sense of control in these environments. Hence, if the parent and youth are not participants within the regular flow of contact between the people and these places it affects how they live, learn, think, and respond to the people and the network as a social system. The capacity to take action is enhanced through Progressive Investing practices, which produce the energy to inspire the parent and youths change through a green learning scheme that empowers the exchange of their labor.

What is the Environment?

The word environment is linked to being human, social, and ecological where there is any contact between self, other people, and their surroundings as environments that interact within the context of other systems. When a person is becoming a human being in the womb, they are an environment within the environment of another person (parent) within the context of other networks. The contact and interaction is reprocessed through sense and receive paths responding to the parent and youths act to participate in new contact and interaction.

The environment provides the means for overcoming four important human information processing needs: a base and space to live, learn, think, and respond to goals. Regardless of who you are or what you are, the earth provides a way to meet these needs when the learner is open to input from all sources. In this sense, the environment is a green learning system that uses the earth to encourage exploration as a product or service to all (Slaton 2009).

The environment is comprehensive as a natural non human world, and complex as a world of living things that include human beings, and effected by man-made things. We are concerned with the natural world of non human environments, because this is the ecosystem that supplies the resources that allow us to build and sustain life on earth. While this is not the area of our focus, it is the essence of our call to action. Helping people learn how to live on earth, which encompass land use and man-made homes, schools, neighborhoods, and workplaces is our focus.

When we use the word environment, it is an inclusive term that may involve multiple factors of experience. This is because the non human world is all around us. We live in a human world of living systems, and we use man-made things to make changes to the natural world. This last point is important, because we use homes, schools, neighborhoods, and workplaces to make changes in the natural non human world and in the world of being human.

For example, homes, schools, neighborhoods, and workplaces are set up by humans for other humans to experience; personal, academic, social, and occupational changes to compete for wealth and resources; in natural and mechanical environments where they must cooperate to participate in the pursuit of success. This is why learning how to live, learn, think, and respond to contact and interaction makes up vital connections between self, other people, and the environment. In order to participate in the pursuit of success, you must have the capacity and potential to learn how to cooperate.

Different environmental experiences set up different issues, problems and concerns for parents, children, youths, and young adults to live through and to address. The experience of living in a highly structured suburban home as opposed to an urban less structured home. The experience of learning in a suburban well funded school as opposed to an urban poorly funded school. The experience of thinking in a well served neighborhood as opposed to an urban less served neighborhood. The experience of responding in a skilled workplace to compete for higher wages as opposed to a less skilled workplace to compete for lower wages set up different test.

Parents have to know who they are from the inside to the outside, and have a sense of who their child is from the inside to the outside as environments. In this sense, the parent and child are self-contained environments. Learning how to be a parent starts with the act to learn about the inner needs of self as a mental function to learn about the brain and the roles of the central nervous system; as environmental functions. The child's brain is in the body as a separate environment for the parent to experience learning about these mental functions. As the parent lives to learn how to study their child's inner feelings, they learn how to work with their own developing sense of self, their child, and the world.

Being a parent is a vital act of environmental leadership. The space, the self, and the child have to be structured for the task. This is why the parent's role is vital to the child's act to live. The parent uses the home; uses the school; uses the neighborhood; or uses the workplace to move the child through these environments to learn about their functions. It is a strategic process of learning how to live. Hence, the Progressive Investing

Model below shows that the certain aim of the practice is to learn how to observe, listen, learn, help, and lead.

Human Systems Model 5

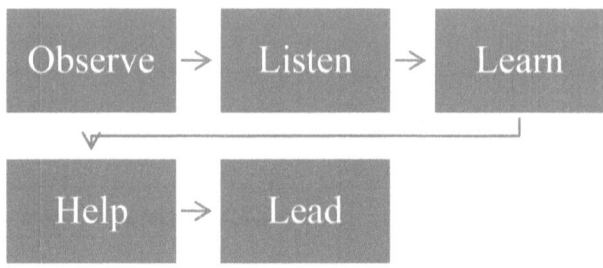

The act to lead starts with self-awareness that comes from learning how to observe, listen, learn, help, and lead other people to learn about their own environments. The environment of self and the environment of the home are the parent's environments of leadership that sets a child up to experience the needs and difficulty they will face in real life. Learning how to go to school, in this case, is a real need and a difficult task the parent and child will have to face together in real life. The parent will need to learn how to set teachers up to experience learning how to learn about their child, and in relation to the environment of self and school.

To do this, the parent has to share their practice of learning how to observe, listen, learn, help, and lead their child to go to school in the right state of mind as a thing leaders do to help other human beings learn with them. Right away, the teacher will learn that while the parent may value academics, the most important thing they look to learn from their talks is that their child's personal growth as a human being is just as important (Armstrong 2006). As a parent you want your child to enjoy going to school to learn how to pursue their right to learn how to dream about the real things in life that people say matters most.

Images of the best homes, the best schools, the best neighborhoods, and the best workplaces are to be pursued by all parents in their role to lead and learn from where they are in the man-made ecosystem. With this in mind, parents can encourage teachers to think about the role being happy plays in how a child tries to learn in school and live at home. The relationship between the parents role to be a good leader for the child to follow and the teachers function to be a good leader for their student to learn from is framed by these needs and values. The teacher has to be led to feel these values as the things that protect a child's right to dream of becoming the leader of their own personal and academic environment someday.

As a parent, you want your child to learn how to live and think of social success. Being successful requires a desire to learn how to live with self, other people, and their environments. This means your child's sense of feel for where they belong has to be set up for them to experience. The neighborhood is the social place where a parent can set up events to help other people learn how to help them lead their child to become more connected. Feeling connected is vital to a parent and child's success with academic and social events.

"Each time we learn something new we are having fun (Sullo 2007). To learn how to have fun helps to motivate the parent and child to learn how to live with self, other people, and in their environments. The neighborhood is where being motivated or driven to live for and with certain values come to the forefront. Your sense of self, other people and the world shows how connected you feel to the right to dream and pursue a happy state of mind to become a leader, helper, and good neighbor.

These are the things you learn from within your sense of feel for self, how to control your external feel for the world to think. To be happy you have to live with a sense of personal, academic, and social control which reflects your capacity and potential to lead. In the first task you have to live to learn how to feel happy. To feel motivated; is the second task, to learn how to learn to improve the academic skill sets you will need to talk through as a family leader. The right to feel the dream and to pursue these goals in reality; is the third task, to think through the relationships between self, other people, and the environments of environments.

This is a system of strategic tasks. To observe means to sense—study, examine, and detect—an effect. To listen means to feel—hear, record, and experience—an effect. To learn means to focus—contact and interaction to cooperate with an effect. To help means to center values, goals and motives in the effect. To lead means to participate—set up, structure, and apply—systems and strategies to control and manage all effects. "In the human systems age, it is the point of human systems research that these functions be recognized, since the concept of body and mind (brain) are used interchangeably with being physical and mental, to form a distinct method for the control of emotion" (Slaton 2009).

Teaching and learning from this perspective is, hence, a new human science and flows from the study of learning, behavior, and conduct that is based on personal, academic, and social responses to the environment of environmental experiences. For instance, we are in a discourse about how personal issues affect behavior; how academic problems affect learning; and how social concerns affect conduct. When the brain as an internal environment, the body as an external environment, and the senses as internal and external networks of the human body are not set up for the experience of contact; the physics of being mental, social, and environmental, can be found to be at odds with the basic flow of energy between these and other factors that affects the whole human system.

The Progressive Investing Model below describes the main ideas behind our efforts to improve learning and support services. As you improve the parent and child's capacity and potential for learning, you improve how they receive academics. You improve how they control their behavior to process the issue. You improve how they sense, feel, and focus to respond to concerns.

Human Systems Model 6

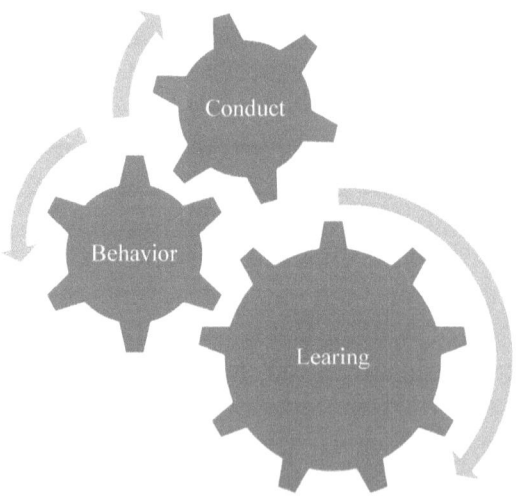

Learning disorders are diagnosed when the individual's achievement on individually administered, standardized tests in reading, mathematics, or writing expression is substantially below that expected for age, schooling, and level of intelligence (DSM-IV-TR 2002). Learning problems are assessed when the child's capacity and potential to receive or to learn from instruction how to read, write, do math, speak, listen, and act are significantly affected by personal, academic, and social difficulties in the sense, feel, and focus process cycle.

In other words, we use the sense, feel, and focus process cycle to assess the child's capacity and potential to learn and receive instruction as, and from multiple environments. The way the child sends feelings forward is an expression of internal control, and the way the child receives thoughts backward is an expression of external senses of order. If the child does not maintain a calm state then the path to send feelings forward is muddled and the path to receive thoughts backward is less recognized.

Hence, there is a breakdown between the initial contact (parenting) and the subsequent interaction (teaching). You assess the child's sense of self for internal tension; the child's sense of other people (parent) for external stress; and the child's sense of the environment (teacher) for external pressure. If the child cannot feel contact from the inside out, then there is a pattern that describes a lack of awareness. This is called a behavior pattern.

When you study a behavior pattern it is to learn what is in the child's act to live in relation to home and a personal sense of self; learn in relation to school and an academic sense of self; and think in relation to neighborhood and a social sense of self. We take these steps to learn from the person, where they are in the green learning systems of self, other people, and the world. As we have pointed out, this is a comprehensive and complex process cycle. "The problem is that by creating an environment during the first 12 years of schooling that focuses on education not as a means of personal fulfillment but as a way of obtaining high test scores, we condemn many kids to school frustration and the subsequent abandonment of school altogether" (Armstrong 2006).

When learning problems go unaddressed they affect how the child learns and behaves in social places. The home represents the forming identity of the child, the school represents the forming ego of the child, and the neighborhood represents the forming superego of the child; in this sense, because that is where the youth or young adult will drift when they are in crisis; to think. This is done to withdraw from stress at home, to disconnect from pressure at school, and to search for emptier, less controlled and managed space.

If we follow this logic, a learning problem may be created at home or at school but in relation to the child, youth, or young adult's personal sense of self. A poor behavior is a result of having less of the capacity and potential to display personal and academic skills when asked to participate in home or school events. Hence, a conduct problem will describe the child, youth, or young adult's attempts to hide, cheat, steal, or use aggression to ward off less controlled signs of tension, stress, and pressure; through the use of perceived; subculture values.

This child's life dangles in the balance, because they may not know how to sense, feel, and focus how they live, learn, think, and respond to self, other people, and the environments of environments. This is why the parent has to learn how to practice leading, to increase the understanding, communication, respect, and cooperation between their child and the teacher (Boynton and Boynton 2005). If there is a deficit in the sensory system, the parent is better set up to share experiences of living and learning with their child to the teacher, and listens to how the teacher responds.

The system of schooling will not meet the parent and child's real time needs; unless the parent is prepared to lead the teacher to learn how to teach to their child's brain and not to their child's body. The result is that we find educational practices of today frequently showing confusion of conflicting trends, often in the same system, and the so-called progressive and dynamic tendencies employing concepts and practices inconsistent with their basic beliefs (Taba, 1932). Hence, the environmental learning process has to be expanded to include the human aspects of sensing, feeling, and focusing to live, learn, think, and respond in comprehensive and complex environments; not connected in the totality of its activity and experience of the child as a human learning system.

What is Emotion?

Emotion is the energy that results from feeling things that influence sense and receive paths. When a person has control over the human system, the brain, body, and senses are set up as a learning system to manage the flow of feelings to and from the brain. The potential for an open or closed state of control is influenced by the way sense data makes contact with the body, and the way the brain receives data as energy. This means, emotion moves the potential to change a steady state to a less organized state that reduces the capacity to process feelings into thought, produced by the brain.

When the parent or child in crisis is able to organize how he or she uses the senses to feel and focus on becoming mental, then this is because they are able to transfer emotion, to change the way feelings flow through them. The transfer of emotion is an act of awareness and a process for moving the analysis of contact to the brain through feelings of thought (Slaton, 2009). Hence, emotion is overcome through the process of thought.

Emotion is what is in the forefront of every brain, body, and sense experience. In children that have been hurt by major life events, the emotional state is related to their contact and interaction in home, school, and neighborhood, networks. Hence, our approach to the study of emotion is to use words to set up a method to look at possible causes as events, and effects as signs of agitation, and possible affects on the child (Slaton 2004; Slaton 2008). The task facing the child in the experimental context; learning through stages of school failure; is as a rule, beyond the child's present capabilities and cannot be solved by existing skills (Vygotsky 1978).

For instance, Bobby cannot stop talking about how bad going to school makes him feel. Bobby said he cannot do anything right, and the other kids hate him. He cannot hide his anger. When the teacher asks him to read, his eyes tighten as if to say don't call me out. I cannot stand school. Bobby said the poor grades he gets make him want to be somewhere else. Going to school is not working out for Bobby.

The experience of school in a state of disappointment was too much for the child to overcome without the development of new skills. In this case the objective was to use home and school visits to learn about the child's experience of emotion and to frame a cause and effect human activity system (Slaton, 2009). We were able to frame academic failure, perceived lack of ability, and perceived unfairness as causal factors. The causal factors were related to observable signs of emotion such as anger, fear, and anxiety through charting the frequency of facial and body expressions. The affect was set up using the following terms: personal growth, social behavior, academic learning, and environmental development; to assess these relationships.

The Human Systems Model below reflects the way we framed the flow of emotion in connection with our aim to improve the delivery of learning and support service.

Human Systems Model 7

As we improve America's schools, we will improve the quality of life in America's families. "Too many Americans are just holding on" (Hartman, 2006). Hurt by divorce, alcoholism, lack of efficient skill sets; unaffordable housing, groceries, and utility costs; and poor access to good home, school, neighborhood, and workplace networks are major life events; signs of crisis.

Through this approach we are able to use home, school, neighborhood, and workplace visits to assess the brain, body, and sense experiences in these places. We call our method evidence-based reporting because it helps us to objectify inner mental processes; external stimulus-response effects, and follow up on physical signs of emotion (Vygotsky, 1978). We were able to show the child and talk to the parent about how they prepare for contact in the home and at school.

In the process of setting up authentic learning schemes for the parent and child to experience, we were able to talk plainly about the problem of academic failure; concern of perceived lack of ability; and perceived unfairness. This is how we were able to relate these variables to signs of anger, fear, and anxiety as a family of emotions. Through Building Human Asset Meetings with parents, teachers, and school principals we were able to draw from the words they used the most in describing a list of possible affects.

We related personal growth, of course, to the home and the parent's concerns. Social behavior was related to both home and school, and parent and teacher concerns. Academic failure was related to the child's fears and the teacher's concerns. Environmental development was related to concerns involving human relations. In all these cases, we believe we improved home to school relations by structuring these physical and mental events to revolve around the child's need for new skills.

For instance, we chose to only actively work with a parent and child for one school year, and to use parent reports and school records to chart the development of interaction and thinking skills. What is crucial to note, after the first year we would not attend any more student study team sessions, individual education program meetings, parent teacher conferences; or meetings with social workers, mental health workers and so on. All problem solving help was done through telephone conferences with the parent and periodic workshops with the child, youth, or young adult.

We learned that our new approach to the problem was scientific, because we were studying how the child receives, processes, and responds to contact, which in turn led us to new methods to study and assess their emotions. For instance, human systems research recognizes that the environment can act as a stimulus for a response, but that the brain plays a systemic role in any bodily response to contact. This is because the sense, feel and focus cycle is active upon contact, which means a direct feed to the brain exists.

What the receive, process and respond cycle tested was the flow of energy between points of contact and interaction. For instance, as the child is given an instruction we would record how they handled being told to do something. We learned from these observations that the child was making choices. This meant the emotional disturbance was being supported by an internal event that caused certain levels of resistance. For instance, it may have taken the child three to five seconds to stare us down, and another two to physically respond to our contact through the instruction.

On the surface it may appear as though the child was not processing the instruction, but when we looked at the response, there was thought behind it. We learned this through follow up questions. Such as, how did that make you feel? We described the child's response process as a feeling system (Slaton 2002). The child does not feel the emotion feeding their response forward, but once the response becomes social, the child can feel our responses backward as feedback. When the child is sent a new instruction, we are able to learn more about the state of their mental processing. Hence, there were signs of change in the response that reflected recognition of us as a stimulus. The Progressive Investing Model below describes the sense and receive paths, and process cycles that move the flow of emotion to the brain.

Human Systems Model 8

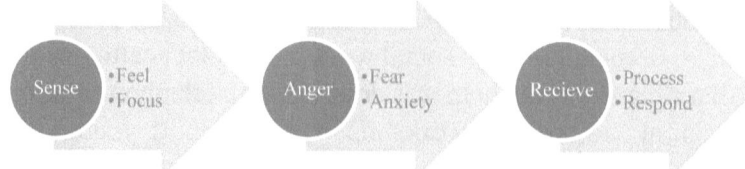

Emotion is in the body, and our contact with the child's brain is physical. Hence, we were being emotionally received by the child's senses. Human systems research recognizes that the study begins with the self, and through contact moves to another level of arousal. Hence, the act to receive the contact is affected by the child's state of mind which is agitated; as signs of anger were noted. This means the senses are being restricted through a level of resistance to our contact at the forefront of the sense and feel cycle.

This is important, because the goal to receive and process our contact also revealed signs of resistance. In other words, when you apply the amount of time it took the child to release a response, and look at the frequency of these cutoffs, you in turn learn more about anger as a block and signs of fear as an act to hide feelings. As a number of different test were run, including reading out loud, writing and presenting, skit and role play. The receive, process, and respond cycle is the point at which the child's brain captures the instruction; and may act to reduce the emotional tension; or the body may maintain control through feelings of anxiety that cutoff the transfer process.

This means the focus cycle carries emotion that has to be reduced through the experience of our interaction between feelings of anger and fear, long enough to process the act to become more informed by the brain. Hence, we look to the content that is in the child's response to our contact, to become more informed. When we look forward, between us and the child, and the child looks backward between their sense of self and us the send and receive paths reveal more about the transfer of sense data to the brain.

From these reports, it should be clear that a stimulus-response framework was expanded; to include process cycles; through these tests to expand the study of how humans learn, behave, and conduct self when the brain is hurt by major life events. The benefit is the qualitative feelings, such as anger, fear and joy can now be measured to assess the process cycles more carefully.

Human Systems Model 9

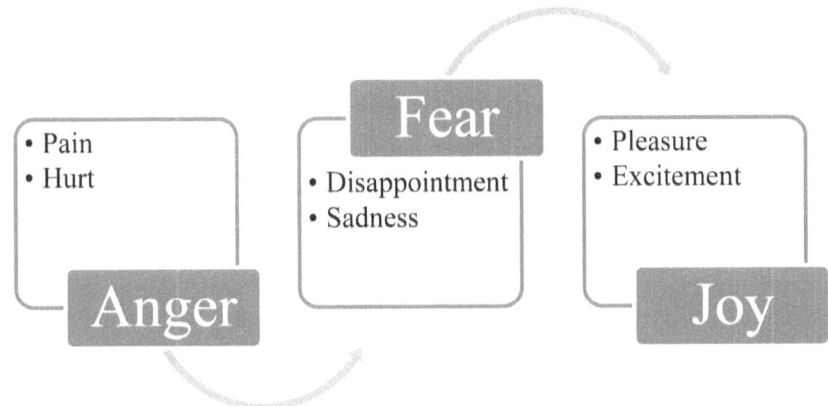

Progressive Investing Model 3

Anger is at the forefront of every experience in the body of a child growing up hurt. It is the pain that has to be processed. To reduce the signs of fear, which cause the child to feel disappointed, goals must be set up to measure the feelings of contact between the researcher and the child. Hence, we used human systems research to chart the child's start point as an event unfolds; an endpoint, as a goal to record signs of focus. This allows us to study the midpoint as an area of affect, to look at signs of emotion as the child and researcher make contact and interact along the continuum.

Each frame is a separate level of activity that we may relate to each other through Einstein's theory of relativity (1947). First we are using a family of related emotions from one frame to the next. Second, we have set up a continuum to help us record how the child feels our contact and how we are being responded to by the child's physical interaction patterns. What we want the child to do is process their fears, by recognizing their state of anger through showing them signs of pleasure; as signs of care.

We are the models, the child is the only participant, and only if they choose to participate in our event. This makes the process authentic. If the child chooses to participate, then their state of fear is being handled as a painful experience of learning how to move toward the goal to focus feelings. This in turn allows us to look at the changes that are occurring in the child's displays of fear, disappointment, or sadness.

When a child finds joy in something it may be a sign of passion. When a child finds pleasure in something it may be a sign of focus. When a child finds excitement in something it may be a sign of processing. The fact is that, each of these experiences is a sign of the child's act to participate: to move beyond feelings of failure; perceived lack of ability; and perceived lack of fairness. We will describe in later chapters how the Building Human Asset Process helps us to link these responses back to the chief concerns of parents, children, teachers, and school principals; more effectively.

Across the action learning continuum of human science, the human systems researcher is experiencing ways to learn with the parent and child through the study of contact that is applied to the use of human ecology. How the land is used to develop home, school, neighborhood, and workplace networks for other humans to experience is explained as a social practice to talk about ways to make sense of the world of business. Human ecology is a human systems science that is open to the mixed sciences point of view, to allow science, scientist, and researchers to live, learn, think, and respond in ways that connect them to the work they do to improve the participant's labor and, hence, sense of the world (Slaton 2009).

What is Sensory Organization?

Sensory organization is the coordination of brain, body, and sense events to move through contact more efficiently. Hence, we ask the parent and child to take action to become more informed from the experience of contact; the parent and child works through the use of their senses to become more aware in regards to the things they see, hear, feel, tastes, and experience as pressure.

You move through language, cognitive, physical, mental, social, environmental, and emotional states to learn how to live with the capacity and potential to feel things. But how do these states become structured in the order they are being presented? You have to make sense of their use. Do they relate to how you might want to move through a problem? You use talk, reflection, identity, processing, interaction, reaction, and feelings, but to feel organized through the use of them, they need to make sense to you.

Sensory organization is the coordination of the senses through systems, strategies, and the use of process cycles. In other words, the senses are what allow people to use verbal and nonverbal signs to communicate language as feelings for thought. Reflecting on how babies enter the world is a simple process. Did the baby send feelings forward into the mother's womb, or did the mother sense and feel the baby backward. At what point does a baby's brain start to sense and feel the mother's womb? Could the baby feel the mother's thoughts? We believe the nonverbal communication process starts during the embryonic phase, and that the senses begin to feel for the mother's signs of care.

Language, as reported above, begins the construction process for our systems feeling approach; which sets up our theory of Progressive Investing as the act to become more informed in relation to a goal; using process cycles to record human systems research. The baby's entry into life forms an initial cognitive experience (Chomsky 2002). Because the baby is physical, by touching the baby, the baby sends these nonverbal signals to the brain as backward feed and the process of id or self identity begins (Erikson 1967).

Verbal cues are used to assess the content in a response for mental elements. This is why; the act to live each day to become more informed is social, hence, we learn how the contact turns into interaction. All the while, the sense, feel and focus cycle is guiding the emotional experiences of experience to become more informed through the act to receive contact; process the interaction; and respond to feelings for thought. We call this a systems feeling approach, because the intent from the start is to learn how the senses feel emotion, and organize states of discipline and self-control.

Sensory organization represents the method we have presented thus far, as a means to coordinate the senses in children, youths, and young adults that have been hurt by major life events. Think through the first step. Why are we open to language? How do we experience cognition? When do we feel being physical? What causes us to become mental? Where do we learn about being social? Who controls emotion? The sense, feel and focus cycle represent a basic processing cycle, where to receive, process and respond represents a complex language processing system.

A child does not learn to sense the meaning of academics. A child has to learn how to receive academics, and how to develop a sense of feel for lines of thought about the use of academic experiences. The child learns how to receive academics into the body by observing how signs, symbols, noises and sounds are used to feel for a sense of what they may mean using the brain. The experience of contact, while in the womb sets the child's brain up for the experience of contact outside the womb. This means that the parent's talking, reading, and writing while the baby is in the womb, can, and may be experienced and experienced.

This is complex sensing and feeling: the mother's body makes up a baby's initial environment; and hence, initial cognitive experience. The senses are set up to feel for objects without having a sense of them, or to feel for signs of communication, which means to experience them. The central nervous system is wired as the brain's communications network in this regard. As I send my feelings forward, you can choose to feel them, or you can choose to ignore them. If you choose to accept them, then your brain is sifting through my words to capture a sense of feel for what they mean.

You may want to ask. Does a baby have the ability to choose? Or, does the baby experience the sensing and feeling no matter what? A baby, in the womb, does not have the capacity to choose between good and bad sensations, i.e., drugs introduced while they are in utero. These are just extended thoughts about the complex process of being born out of conditions that spawn from crisis. This area of science through Human Systems Research requires additional study and analysis to more clearly describe these influences as the child moves up in age.

Acts to interpret words are acts to live through them to points where you experience lines of becoming focused. This is the brain's way of organizing sense data to manage the flow of feelings throughout the body, as signs of control. Through signs of control, we can observe discipline as a physical state. A physical state of discipline is a cognitive object that can be experienced in the minds of self, other people and their environments as sensory organization. The senses are disciplined, the brain is a sign of control, and the body is a physical object to experience.

Feelings are being managed through the social act to live in relation to the experience of contact, and to learn through the practice of interacting to focus how you think through the experience of the environment.

Human Systems Model 10

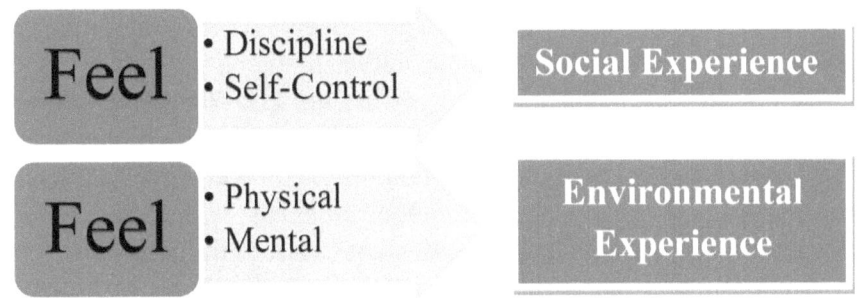

The relationship between the body and the brain is the social experience of sense data and the environment as an experience of focus. Based on how the contact feels as the body moves to interact with the external and internal experience of focus as signs of learning, emotion is reduced backward. The social self is controlled through higher mental and physical states, which open cognitive structures to learn from things that have been emotional to experience, and to feel for signs of care.

When the brain has been hurt, the body (cognitive structures) has to be set up, to feel things backward. The goal to enjoy things, for instance, requires signs of care through the use of talk to change this effect backward. It is easier to feel for signs of care in the words people use in talk. They have to aim the words they use to appeal to your brain. You feel words for signs of care. Hence, you feel for things that will help you sustain a line of focus to act through the contact. All the while, you are learning how to interact with your brain to think through the problem of hurt.

You use your senses to learn how to set environments up to be experienced through the way people use talk. The human system, you have now built up, responds to signs of care. Words that cause or raise concerns, will enter your system as signs from a person who lacks a sense of care for how you learn to receive information. In order to receive words, they have to show signs of care for how they are designed to feel for your brain and body.

The senses are organized to help you live through signs of care. Signs of care are what your brain looks for in contact. If there are no signs of care, the brain enters a protection mode to block and resist the internal drive to interact. If the bad experience continues, this is where the cognitive blocks create internal problems for your information processing system. The sadness will reduce your act to receive contact, seek ways to interact, and search for signs of cooperation.

In other words, cognitive structures are either open or closed based how you experience these environments of environments. When your senses are not being allowed to send and receive (control and manage) the flow of energy, it reduces your capacity and potential to participate. If you are not able to participate, then you are not able to practice sending feelings forward and receiving thoughts backward, to build a higher senses of self, other people, and their environments to compete. The Progressive Investing Model reflects the emergence of the system, strategy, and process. We will use these signs from right to left to think through what we have said about each stage; language, cognitive, physical, mental, social, emotional, environmental, and sensory states.

Progressive Investing Model 4

You are competing to change with states of becoming human. Each sign is a separate system, strategy, and process cycle in learning how to live from the inside to the outside through brain, body and sense events. You recover from agitated states of anger, pain and hurt by improving how the brain receive thoughts backward to focus the body and send feelings forward to sense and feel. Signs of joy, pleasure, and excitement are natural human aims of energy that offset the subjective states in feelings of fear, disappointment, and sadness.

Language expresses how you feel. An attitude for instance, connects cognitive, physical, mental, and social states to the environment of environments; such as through being angry or excited. You learn through sensory organization, how to work through a sense of feel your own behavior to focus and live through the way you learn to display personality, character, and then attitude as an emotional state. In other words, you learn how to sense, feel, and focus feelings.

What is Green Technology?

Green Technology is a term used in reference to the evolution of the person as a functional system, human system, and hence, reprocessed technology which reflects a use of energy, control, and reprocess cycles to become more aware through brain, body, and sense coordination. Our new human system theory and practice seeks to reduce the gap between functional and dysfunctional states of labor for parents and children in crisis; due to poor school experience or from being upset by major life events in man-made home, school, neighborhood, and workplace networks.

Hence, our aim has been to learn how to reprocess good feelings. We want to learn how to help the parent and child learn ways to reprocess good feeling from old, static, and bad feelings. The focus, then, is on how the parent and child learn to move through the process to grow, mature and develop from the experience of feeling things. We have learned that information processing skills, social knowledge, sense experience and reflective feelings of thought enhances the parent and child's capacity and potential to respond to self, other people and the environment.

Basic words—*sense, feel,* and *focus; receive, process, respond; live, learn, think, respond; help the helper help; aim, care,* and *feel*—relate to how the parent and child learn to move through the crisis. Easy words to receive and make sense of are used to change bad feelings that have had a negative effect on cognitive structures shaped from poor experiences. To change a poor behavior however, Amen reports that you have to counteract feelings of negative thoughts in the brain (2010). Hence, words are used to talk to the brain to enhance the practice of learning how to move through poor experiences.

Living each day to learn how to improve their responses to tension, stress, and pressure these words are reprocessed through talk, reading, and writing. Hence, the parent and child has to be signs of discipline and self-control to learn from the experience of experience as sense data and information flows through their interactions with self, other people, and their environments. Emotions, feelings, and thoughts are moved through levels of control.

A green technology is what the person becomes as they are learning how to use a sense, feel and focus process cycle to make better choices and to take better action to improve how they do things on earth to live, learn, think, and respond to change. According to Goleman (1995), this would be a choice to take action to relearn from emotional experiences stored in cognitive memories of pain or hurt. As a green technology, the reclaiming of the brain's body means to be a sign and object of both making good choices and taking action to live each day to become more informed as a human systems science (Slaton 2009).

Instead of living life like a machine preparing to be turned into particles of dust, as a green technology, you live each day to learn how to sense and feel for things to reprocess skills. Hence, your focus is not to allow man-made technologies to dehumanize the person you are as you age, and begin to lose capacity and potential to compete. Your goal is to build up both the capacity and potential to be a good thinker and to reprocess certain sense experiences at high levels to receive input, throughput, and output from observable and unobservable signs and objects.

From the foregoing it ought to be clearer that the sensory organization system is a structure to improve research, learning, and teaching skills through the reuse of feelings and focus; awareness.

Progressive Investing Model 5

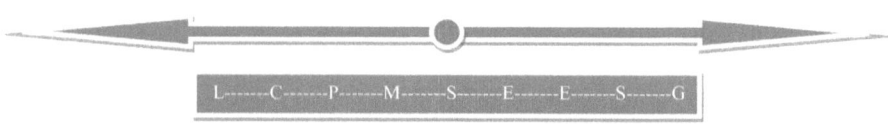

The Progressive Investing Model above adds the words *green technology* as the reprocess cycle in the practice, system, process, and strategy. Becoming a better participant and observer (Stringer 1999), will help you learn how to learn from the study of self and other people; how to reuse language more carefully. What the environmental learning cycle of living does as a more open system, is equip you with the skill to allow feelings to move right through you. In other words, your brain, body, and senses are set up to receive input, process throughput, and respond as output to the environment. This information is reused by the environment to learn how to live with you, as one of its many users.

As you learn how to use talk, read, and write through self, other people and environmental contact you are building a capacity and potential to choose how you interact, store and retrieve sense experience from cognitive memories. Physically, as sense data moves through the brain as a sign of curiosity and the body as an object of discipline, you seek to accommodate the flow of this interaction as a living human system. The mental build up of human energy is released as an act to assimilate with a forming sense of self, other people, and the environment.

Human Systems Model 11

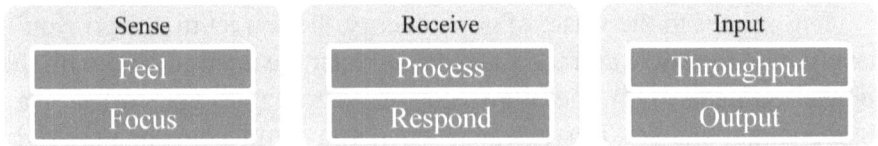

The key block to our structure, is the choice to practice taking action for the good of the earth through social contact and interaction in the process of learning how to sense/receive/input, feel/process/throughput, and focus/respond/output.

Social action requires the strength to be a sign of good and an object of focus, when you are trying to live right and do things the right way. Environments react to signs of care. When our inner spirit is weak, it is easier to live and do things the wrong way, because it takes courage to show signs of care for self, other people, and the environment at the same time. This is why it's so hard to manage the flow of emotion through the body, because bad feelings are meant to be handled by the brain.

Sensory organization skills, as you will learn through the following chapters can be taught. You can improve the way you recall information through the way you identify the cause(s) of anger, pain, and hurt through game play, skits and so on (Goldman 1995). As you are learning how to build up the capacity and potential to live as a green technology, there is an element of mind called judgment that improves with good choices and well thought out action. As you begin to use your time on earth more wisely, action learning continuums allow you to form a practical sense of feel for these skill sets.

The home, school, neighborhood, and workplace networks are places that may be called test sites in the sense that they are man-made environments designed to provoke emotions that create weaknesses in moral judgment through the way one chooses to move through them. This is how people lose capacity and potential to sense, feel, and focus reality as bad choices and actions in these places can drive the body for a lifetime, with no sense of feel for the brain.

We use human systems research to fulfill this basic aim, to help people in continuity with our need to reprocess the brain's body. As natural living systems, we are subject to the nature of change as we form the human skill to change. Hence, we are here to help you change the things you can, learn from the things you cannot, and focus on living as a human technology in harmony with self, other people, and the environment.

This relates to the waste of good energy. "If you let things go, you'll have to pay for later; and the longer you resist the need to deal with the things you need to do, the more you'll have to pay to get started. Wait long enough, and you'll pay dearly; when you could have done the right thing all along at little cost" (Hartman 2006). You have to do the work. When it feels good to learn, save how you feel. When you do not feel as good, reflect on the experience of learning how to feel things. You learn to act through these feelings. As you learn how to reprocess the goal to live each day to become more informed, these feelings grow to be a way of thought.

Chapter 3

Our Thoughts on Family Leadership

Linking Parenting to the Childs Role

Learn how to use the role of parenting to improve your ability to lead and develop a child as a learner and participant in your family system. These words feel good right? Every parent has to learn how to lead and develop meaningful roles for children in a family; and every child has to learn how to be a learner and participant to value what it means to be in a family. These words seem aimed to you? Do you think a parent and child needs to practice learning how to live together?

Can you learn how to live? Do you want to learn how to live with children? Do you know how it feels to live with a child? What does a child feel, when they live with you? Are you set up to lead a family? How do these words make you feel? They are simple words to complex questions: how are you? How do you feel? What is in you? How you respond, depends on the type of person you are. If you are open to learning, then these words will be reused by you, to think through and feel the issue. Can you lead?

Is there a system, process, or strategy, or do most parents learn how to lead a child through trial and error? How do a parent and child learn to work together in ways that foster roles to lead—learn, develop—participate to focus the well-being of family? Whether a parent knows it or not, they need to learn how to lead and develop a child's ability and desire to be a learner and participant in family improvement.

Let's Talk about Family Leadership

We have to talk about family leadership, because we are concerned with child development. For instance, we are concerned with how a parent leads a family, and how a child grows up in one as a learner. Most of the parents who come to us for help are single mothers with children in crisis. We work with parents to help them lead children that are growing up hurt by major life events, to learn how to live in their home, school, neighborhood, and workplace networks. These are the main places a parent has to be a leader in. But they each require certain skill sets.

A parent has to build up a feel for observing contact and interaction through the study of how a family learns to live together. In a parent's role to observe contact, is the task to listen to the words being used in talk? As a parent listens to interaction, they have to be set up to learn from the experience of how they feel. In the experience of learning how to feel the contact and interaction, the parent is in the role to help other people learn how to live with them. The parent shifts from helping others learn how to live with them, in the role to lead them.

In a few words, a family leader has to observe, listen, learn, help, and lead. Skills!

Human Systems Model 12

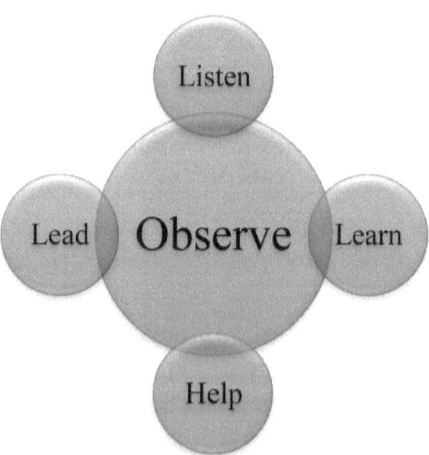

Being a family leader means you have to feel your way through contact and interaction, between self, family members, and their environments. Hence, the purpose of an observation is to focus the senses to feel for contact and interaction. When a parent is set up to listen they are ready to hear things through signs of discipline to control their feelings. This is a form of learning how to learn from the experience of feelings that may move you, if you are not prepared to accept them. These acts work to build up the parent's capacity and potential to help self, learn how to help other people—help. Helping is in the process to become a successful leader, because the parent has to accept the idea, they need to help other people learn how to live with them.

The above Progressive Investing Model frames the practice that works well with our process cycles. You use your senses to observe things. You look for ways to listen, because you want to hear how other family members use talk. At the same time you are listening, you are learning to display forms of discipline through your act to learn from the experience which calls for self-control. You are automatically in the mode to help yourself through the process cycle, as a form of self-leadership. Self-leadership is what you are really aiming for. To be a good family leader, you need to learn how to lead your sense of self, other people, and their environments to feel for you.

Now we can talk about Family.

We use words to solve problems of thought. The advice we have to share has been built up over the years. We learn with you, how to learn from you, to help us help you. If you study the words we use to talk through, you will learn from the science of processing words. Each word is supposed to enter your system, to arouse certain feelings. In this case, a way of learning through a threat or crisis, the words lead you to act like you want to live. We mean you have to act like you want to become more skilled.

Developing parents as family advisors is a method we use to structure the roles and functions of a parent, as a family leader in relation to how they learn to receive, process, and respond to our words through talk. As family advisers, we prepare a parent to act as a family leader. We help parents learn how to act through a poor behavior. Hence, a parent learns how to study the act and process of taking advice to learn ways to lead. This means, the parent learns how to study a family from the role of an advisor.

As family advisors, we want to learn how to help children that have been hurt by major life events in their home, school, neighborhood or workplace networks, work through an attitude that show signs of anger. This means, we are concerned about parenting, which may include being a mother or father; male and female relationships. Hence, how a child experiences family life is extremely relative to this concern. What this means, then, is that we are here to help children—sons and daughters—learn how to move through the K-12 school system while they may be growing up hurt by their own family relations.

For example, when a child is growing up hurt by family affairs he or she may not know how to relate with their siblings as brother or sister when the relationship has been upset by breakdowns in parenting; due to poor patterns of mother, father, spouse, or friend relationships that affect them as children, youths, or young adults. How a child feels about a parent or adult relationship is based on how it causes him or her to feel or not to feel things. Not to be able to feel for a son or daughter; brother or sister is a lack of feelings in terms of care; which means to feel for and think about who people are, and what they may mean to self and others.

When a parent or child does not want to think about how it feels to live together or without each other this is a crisis and major life event. Especially for the parent, because a parent has to show signs of care to help the child come to think about how they feel. Otherwise attitude: being tired or being upset. To help parents and children move through a crisis, we try to meet as many of their close relatives as is possible.

How close relatives: parents, grandparents, aunts, uncles, and cousins feel about each other and their members that are in crisis, is important to us. In this area, we are concerned with family learning, in the sense of how a family learns to live apart and together through their action and reaction to us. This is because a family is a group of people that are related to each other as an internal network. The reason the latter is important, is because a family, like an individual member, may be loaded with internal tension. Hence, as we may move in or out of a parent and child's environment, this may cause anxiety.

However, when a parent comes to us for help, this is an act to learn how to receive help. The way we help a family is to study it first as a community. In our act to learn, we meet with the people that make up its core membership. This is done through community-based action research (Stringer 1999). We have to learn how to help improve the parent and child's capacity and potential to take action to live with how it feels to work through us to look at family issues, problems, and concerns.

Human Systems Model 13

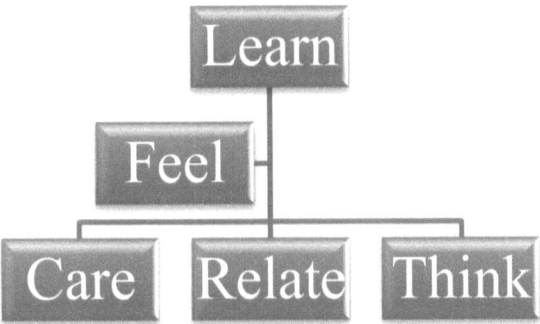

We learn from our act to help, how to feel for the parent and child as a family unit. To grasp a sense of the parent and child in relation to the goal of helping, our search for understanding each of them as individuals and group members is important to us. This is where we learn about how a parent and child may feel things backward as we think things forward for them to experience. There must be signs of care in our act to help, and in the parent and child's reaction to us as helpers. If there are signs of a lack of care, then we use talk in an effort to make it easier to grasp a sense of feel for us as we talk.

In the process cycle to observe, listen, and learn we look for ways to relate with the parent and child as their helpers and for them to relate with us as having some sense of understanding about the experience of being hurt. As we continue to send our thoughts forward, we look for them to send their feelings backward to give us this sense of understanding, how to use our signs of talk. Research-based practices like these are shared with the parent and child to improve how they learn, feel, care, relate and think with us; as their learning consultants.

The research-based process we share with the parent and child is Progressive Investing. We take action to learn how to study the way they seek to live each day to become more informed about their parent and child relationship. We think this helps them relate the things they learn how to do, back to the use of talk. Hence, reality-based data is being used to set the parent and child up to learn from the experience of our contact how to move through the problems being discussed. This is what action research is, a way to help the parent and child learn how to make and sustain contact as participants in their own acts to live.

We talk about the needs of children, but we mean parents as well. If you are going to learn how to recover a child that has been hurt by major life events in the family system, then you have to be prepared to learn how to center the work on the needs of the parent to learn how to deliver help.

Human Systems Model 14

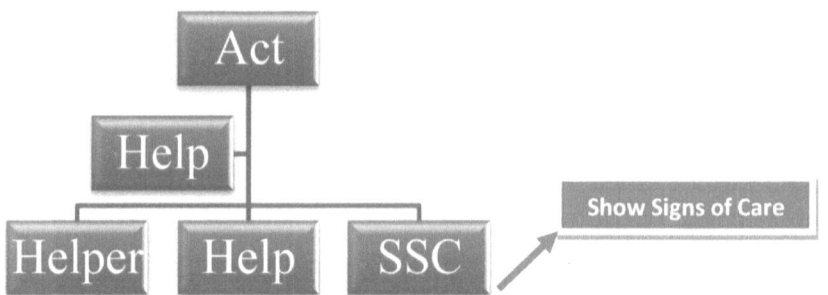

A family is a group of people whose actions are set up based on parenting and leadership skills. As we model the act to help, the parent and child act to learn how to help us. A parent and child learn how to live apart and together through their action and reaction to each other. When a parent or child acts in relation to us, they are learning how to move as a unit. In fact, they are learning how to learn from action. In doing so, the parent and child learn how to relate to each other in relation to us as helpers. Hence, the parent and child become actors in the act to learn how to help. It is through the parent and child's reactions to us, we help the parent lead the child.

When you think about it, the parent and child is learning how to relate to us as team leaders. We call this, learning how to help, the helper help through their act to learn from us. As we move deeper into the process cycle, helping the helper, turns out to mean that the parent and child are learning how to lead us in our work to learn how to help. This is all taking place only if, and when the parent and the child as individuals, choose to participate with us in the lead. That makes the work more authentic. We learn this from the parent and child's effort to "show signs of care" (SSC). To show signs of care, the parent uses our way of talking to the child's brain to talk to us and their child's brain.

Hence, the child's act to learn how to show signs of care lead them to use more caring words to interact through us. The parent and child are learning how to study their own choice of words, as a learning base to feel for the changes in their action toward us. The Progressive Investing Model below refers to this process. The line represents a continuum, the three points between the start of the arrow represent the event, the affect, and the goal.

Hence, we can talk about how a family takes action with the aim to learn as a family, which will show an affect on family relativity or how the family unit relate to each other's act to learn from us. This is what we have been describing, so far. To learn how a family learns requires a method to measure the time, speed, and energy (action) used to react to us. This way, we can make sense of family action as it relates to the pursuit of goals that allow us to describe the affect it has on behavior.

Progressive Investing Model 6

Family Action   Family Learning

•   •   •

Family Relativity

As we meet family members, the issue of relativity speaks to issues of closeness, separation, disconnect, withdrawal and isolation; anger, pain and hurt. This in turn relates to male and female issues, structures and functions, tension, stress and pressure, and educational attainment; values. A family's field of experience is the home, school, and neighborhood, and workplace networks they share. These are symbolic places, where a family learns how to practice living alone and with other people if they choose to do so.

Progressive Investing is a natural action learning process used to focus the action of family members through events designed to guide the act to practice using home, school, neighborhood, and workplace networks more efficiently. In theory, a parent and child live to progress as a team by investing in self, other people, and the environment to learn through community events. We call this community-based learning, too focus action.

Hence, we can talk about the Building Human Asset Process for family leaders to improve the state of family life. This process will be discussed in relation to the crisis of growing up hurt by major life events in a home, school, neighborhood or workplace network. The Progressive Investing

Model below relates the practice to the system, process, and strategy introduced in chapter one. Here, we connect the state of family.

Progressive Investing Model 15

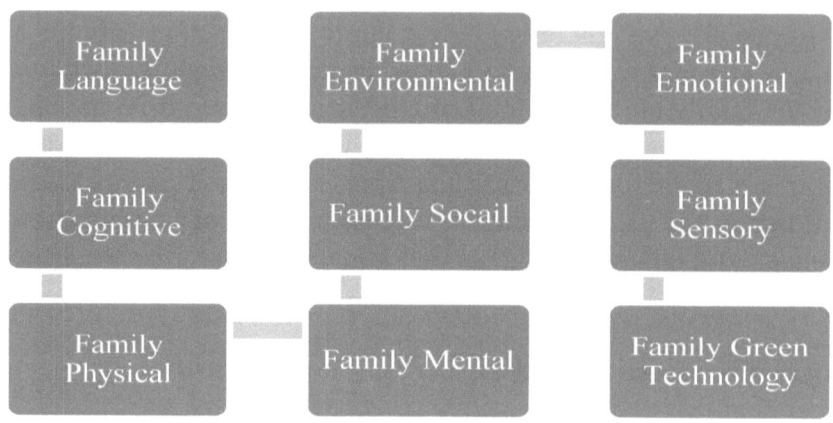

Family Language State

A family will use a certain language style to build on cognitive, physical and mental noises, sounds, signs and symbols to communicate as expressions of cultural relativity and conventions of thought. Knowingly or unknowing, the way a parent's language state is set up will affect how words are selected in the use of talk. Hence, a family's use of language ought to be viewed within the context of survival. In our view, language is an expression of feelings, and we find that language use is guided by circumstances whereas circumstances may control the use of talk.

As expressions of cultural relativity this may mean that poor male and female relations may create high or special need families in crisis, which lack the focus to set up cognitive, physical and mental skills in the ready for parenthood, because of issues, problems, and concerns in which they become social experiences through the way language is used and felt. This may mean that the use of talk is not set up for the experience of care through the use of words. From a sense, feel, and focus perspective, we believe that characteristics like anger, pain and hurt upsets the use of talk.

When you are angry, the body becomes more defensive and less blood flows to the brain as opposed to the arms and hands (Goleman, 1999). How does talk feel, when you are angry? Do you sense your body move into a defensive posture? Can you focus the brain to hear and listen, when you are angry? When you feel pain, you are in a state of anxiety? You can feel the tension in your body as signs of hurt move you to feelings that reflect a state of pain. Yes, you can still hear, but can you focus. Anger attacks the brain, by redirecting the flow of energy. A behavior moves the body into these modes.

You may ball your fist up, you may stare, and you may slap or hit, but how do you use talk? You want to be ready to use talk. The way the use of talk calms you down is through the act and practice of talking to the brain. Allowing feelings and thoughts to flow to the brain creates a process pattern. A process pattern feels and looks like a mental state of focus.

Progressive Investing Model 7

Receive Contact   Respond to Cooperate

Process Interaction

As your brain and body works to sense, feel, and focus the event to receive the contact you respond to cooperate, which is the goal state that affects the process of interaction or the mental state. When you are living through states of anger, pain, and hurt, you have to focus to become mentally aware. You use talk to calm down. To feel the contact, to focus the aim to cooperate, means the senses are being used to build up a mental state of care. You have to show signs of care to move through these feelings as a line of focus.

When you focus how you feel, there may be other underlying characteristics like being poor, being a single parent, and being an absent father as conditions being experienced with bad feelings. Any signs of care may be minimized or blocked based on the way these experiences are being felt through the use of talk or the lack of talk. It is the brain you aim your talk at and not the body. Everything you say and do through the use of talk is relative to the family values you set up to guide your action. A child's body is what you see, which is why you feel for the child's brain. Hence, words are used to talk to the child's brain.

The way you do this work is by looking at the features in your dialect. The way you speak through tones and attitudes may differ from family to family based on the individual choices being made in relation to contact, interaction, and intent. This means how a family in crisis learns to use talk has to be set from one member to the next. You have to implement clear signs of care for self, other people, and their brains to get them to share these values. The way you speak and the tone you use to share these feelings will have an influence on the cognitive and physical capacities and potentials to improve how words are being felt.

The tone and pitch of your voice is set up by the way you deal with the circumstances that surround how you sense, feel, and focus to speak, which reflects on the state of language use in your family. To reduce these threats, you need to be a sign of care to release a caring state of language use. When you get tired, you are already in this state. When your child is agitated, this state of choice is already set up to release a voice that ought to calm them down. This is a sensible way to feel for states of calm and to set up a voice and recognition tone to calm down other family members.

We are talking about the needs of children who live in broken families. You want your child to learn how to use talk at home to build positive feelings of thought. When you move from the cognitive toward the physical experience of talk, you represent feelings of thought to your child's brain. Your child stores images of you talking to them that they use to build their own capacity and potential to think with you. This is another reason to aim your feelings to show high signs of care toward your child's brain through the way their body responds to your choice of words.

Children that live in broken homes need a sense of meaning to live through the experience of how they feel as they grow to realize there may be a problem. This is where word *choice* can really make the most difference, in how your child uses talk to express how they feel. Words enter the body and create calm or disorganized states. The words you use are the words your child will feel to talk through these feelings. If the words you choose to use are confusing, then your child's capacity and potential to make sense of their meaning will seem disorganized.

A child can feel the things they say, and hence, build up a sense of what the words may mean by the way they move through their body to their brain. These feelings lead your child to think the way they feel. So you need to understand that the meaning of your language system is to show signs of care, which means to improve your child's feelings through thoughts of care. A child has to have a focus point that allows them to experience you as their model. If a child can feel your words moving through them in a state of calm, then they can focus the experience of your words.

The experience (your voice and tone) of experience (your feelings) is the experience (your thoughts) of meaning. The speaker (you) experiences the goal (to be a sign of care) to experience (self-improvement) what it means to care, which is the experience (feelings of thought) to aim the brain. The focus of these language experiences is to set up the act to speak through the choice of words with the intent to reach states of contact and interaction. The action between the parent and child's brain and body expresses how these feelings are being stored as memories.

Family Cognitive State

We will continue to work from the initial cognitive state (Chomsky 2002). The child and parents initial cognitive experiences are reviewed to set up sensible ways to improve the systems, strategies, and processes needed to recover parents and children that have been upset by major life events in their home, school, neighborhood, and workplace networks. If the parent and child were in crisis before, during, or after birth, this set up means to start from where the parent and child are at the time of their interactions with us. Meaning, the process to recover starts from where you are right now, in your own problem situations.

We are free to make choices, but we make choices based on self-reference (Wheatley 1999). If we have set up a system of care between the parent, the child, and us as their helpers, then change results from how they comprehend these meanings as they relate to them. Meaning is created in the process to observe, listen, learn, help, and lead self to feel the goal to change. Hence, the parent and child will only choose the goal to change, if it leads them to comprehend the meaning of who they are becoming in the problem situation. "Show signs of care."

Progressive Investing Model 8

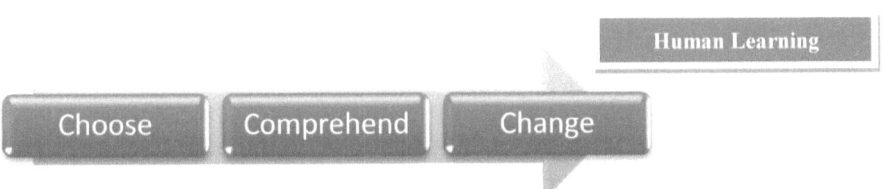

You have to choose to comprehend the need to change, because as a human system, you need to choose to learn how to accept help from us. In the sense of a family cognitive state, you need to move yourself to reflect upon the reasons why you are here. This is what may allow us to talk through the structure of these memories to learn how to move our talk to include learning how to act through the way we may cause you and your child to feel. However, if we continue to use sensible signs of care, this ought to make it easier for you and your child to work through our process cycle to deal with these feelings.

The initial event is for you to choose, which in turn sets up the goal to change with our contact and interaction working to move you. In this same sense of caring for how it may feel to ask you to discuss the past, we aim for the affect to result in us, coming to learn how you comprehend the things you say. Hence, this is the phase where we are here to observe, listen, learn, help and lead only if and when you are in the ready, to follow. This is your memory structure, and we build on our contact and interaction through the use of these words to talk to your brain, and your child's brain. We must realize that either one of you may be closed or open to our contact or interaction.

We can talk to a child who lives in a fatherless home, if they choose to work through you to set up how they need to learn to feel us. Your child is more likely to aim those feelings through us, at you, but to open the mind to the use of talk to feel them. This is a reflective learning cycle.

Human Systems Model 16

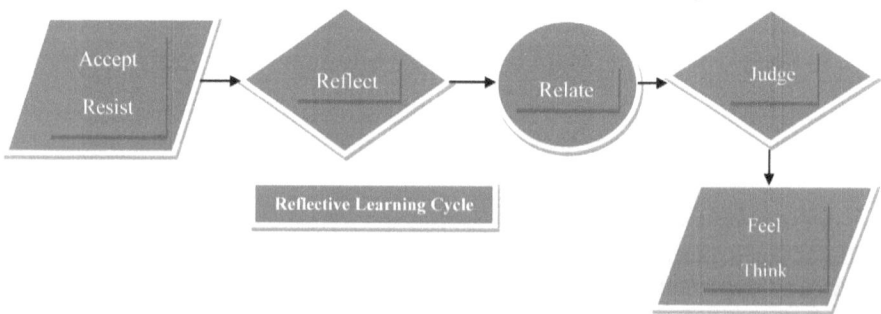

Everyone moves from the event to choose, to the goal to change and to the affect on their capacity and potential to comprehend the human learning systems model as a need to aim the brain. Hence, when we set the parent up to learn from us and the child up to learn with us, we have constructed a reflective learning cycle for them to work through us. The parent or child can choose to accept or resist our contact or interact with it to open the path to reflect on bad experiences, but relate with them through our signs of help. Both the parent and the child will judge the experience based upon how we lead them to feel and think, or not to think and feel.

When you live in a blended family, these memory structures can stunt a child's sense of self to feel their way through the things that make them think differently about a family's cognitive state. These memories can be hidden behind a pretense that they do not exist. But when a child lives in a family made up of extended membership groups with different biological settings than they have, it encourages them to wonder who they are or why they are here. When you are a child growing up hurt, the question dealt with the most, is if anyone cares.

Children that have been hurt by major life events in their home, school, neighborhood, and workplace networks need to be grounded in a clear sense of who they are, to help them make sense of the person they are becoming under those circumstances. When a child is hurt they will bounce between feelings of self-hate, and feelings of self-love, when they are not set up to deal with feelings of shame as they live in search of a reflective state for them to feel and think through to learn.

Who am I? Why am I here? As you set out to explore the answers to these questions, it either has signs of joy, pleasure or excitement in it or fear, disappointment or sadness in it. When you do not know, what you think you ought to know, then these are the feelings being moved. Are you a shame of self, when you lack an understanding about for instance, how your father feels about you? People seem to always be seeking to know a person through who their parents are, or are not.

When a child questions who they are or what they are, they are asking to be shown how to feel, as they move between wanting to learn and wanting to know the difference. Love may be expressed through feelings of joy, pleasure, and excitement. Hate may be expressed through feelings of anger, pain, and hurt. As the child moves to learn about these feelings, shame is what a child may feel when they do not know the things they think they ought to know. Who am I? What am I?

What this means is that they are physical, and can observe, listen, learn, help and lead themselves to feel and think through a state of hopelessness because they lack the capacity and potential to work these circumstances out through their contact and interaction with self, other people, and their environments. This is because everyone has to talk to this child's brain, since they have a different feel for being physical. For instance, they may need to learn how to accept self as being different from others, but through the use of words that leads them to feel and think about who they are and what they are becoming with signs of love. In other words, they have to be set up to feel their way back to clearer cognitive states through the experience.

You talk to each other, to improve the way you think. The way you come to think with each other's brain in mind, changes your behavior, which changes the way you feel for your child to think. The way this changes the behavior is through the act of acting through your negative feelings to perform for your child to sense, feel, and focus their act to live through you. Hence, the practice of talking to your brain with your child's sense of self in mind enhances the experience of positive thinking (Amen 2010). You talk; your child and you experience feelings and thoughts forward and backward through signs of care, experiencing experiences to overcome poor experiences.

When you cause a child that is hurt to reflect, it is to help them realize they have this choice to learn how to feel and think their way through these signs of care. This means the child has to be an open symbol to receive the sounds that help us to observe, listen, learn, and lead them to feel the words of care. Now we can look back to the use of language, as a process that helps the child respond to the meaning of our act, which is to be a symbol of thought and feeling. With you, the parent as the leader, the whole process is being set up to get your child to physically participate in family affairs.

Family Physical State

Now we can talk about the issue of gender. Why don't males and females relate to children, in the same way? Do men feel for children? Being a woman is not the same as being a man. Do men resist the need to feel for children? A male is more distant, more aggressive, and therefore more overtly severe (Blankenhorm 1995). A woman is set up to feel things, but a man has to learn to feel things. More so than not, women feel for ways to accept feelings, men think of ways to resist feelings. Are men more apt to think before they act, or act before they feel things?

Are women more apt to feel things before they act, or act before they think? In other words, the greatest difference between males and females is in the way their brain, body, and senses are set to feel or think. Men need to learn how to feel things because they are not set up from the inside to the outside, the way women are, to feel things. According to Blankenhorn, "Even before "symbolic" consciousness of gender" begins to emerge, "the father is experienced in his total physical and emotional behavior as the exciting, stimulating, separate other" (1995). Males do not feel for, they think for the child. When you send thought forward to feel for the child, you are looking to feel things backward.

The male thinks to feel the child to build up a capacity and potential to accept these feelings. The males physical lack of attachment, affects how males and females think and feel through their roles. Men are more apt to be less reflective than women; hence, they are less likely to relate to how they feel. "Where a woman is nurturing, a man is challenging; where a woman is compassionate and emotionally indulgent, a man is fairly strict and stern, where a woman doles, a male prods" (Peterson 2000).

Women need to learn how to work through the things they feel. Why are so many men, living outside of the family, because they have less feelings for the children (Blanenhorn 1995). Nothing affects a family's physical state more than the absence of fathers. This is to say, the biological and social connection to the child is cutoff by a confused males sense of identity. This automatically upsets the physical state of family, because males are more apt to flee from their feelings of care for children.

Do boys feel for a man's voice, or do girls. What happens when a man gives an instruction to a child in contrast to a woman? Is there any real differences? If a child is fatherless or does not know who their father is, do they search for a sense of his voice through their reflections of their mother's speech acts? Do mothers find themselves trying to speak in multiple tones and pitches to account for this circumstance? It is because we are physical, that a child may grow to feel less appreciated, when one parent is absent.

It is not always that the child does not like being told what to do. Sometimes it's just that a mother's use of talk, can serve as a reminder that the child's father is absent from a physical capacity and potential to have contact and to interact to learn how that feeling would feel. Not knowing how a parent's voice may feel is a problem of thought. You can only imagine the experience of being told "I love you, feel you, care about you, or think about you." Physically, a child may live in search of these feelings.

A family's physical state is based upon how you as a leader use talk to move each member to feel for a sense of self, and to talk through a state of becoming more knowledgeable. Family members need to be signs of openness to move through the problem of learning how to feel things that cause them to think differently. For instance, being different based on race is a moving experience. You can feel how people look at you. You can sense how it feels to be different. How do you move through feelings of being different, when your parents are not together, "long enough" to help you learn how they work through their differences?

When the physics that make your family stand out are mixed, unless you know how to use talk, their feelings may be mixed up. Learning how to live in a family is just as important as learning how to use talk. When a family's talking system is open, members will say things that lead to hidden feelings about how it feels to be in a family. A child will tell you what that looks and feels like, to be a member of your family. Children will tell you what type of family they think they have. These are feelings, children send forward to learn how you feel and think as their physical family leader.

Does race matter to you? What does it mean to feel and be different? Is being different, a thing to talk about? Race as an event, will always matter. How do you move a child through an event? You aim talk at their brain, to help them feel through the words you use to show signs of care. You and your child are signs of how a family feels and think things through to learn how to help. Being different is not an experience that can be easily explained, because you have to do the work to move through it. As you make the effort to move through issues of race, you learn how to talk about it.

Every day, for the rests of their lives, your child will look for a sense of who they are. You want your child to do this from the inside to the outside of their body. A child will live through a sense of you, a sense of self, and a sense of the type of body they have. This is because a family has a physical state. The way one member looks at another is telling, in terms of how they handle being different. We are describing a physical state with mental ramifications, if a parent and child do not learn how to move through their differences. We are saying that to be different, is a physical and mental experience of feelings that need to be channeled.

A family's health is always at risk, when there are family members who do not feel good about who they are, or what they can become. How do you look at self, when you cannot look at family? Family has to have a sense of order that flows from the use of talk to learn how to feel things as a group. When one member falls behind, it threatens the group or family. To fall behind means to lose focus. To not feel signs of care, is a loss of focus. To lose a sense of who you are, and what you can be, is a lack of feel. Being physical is a skill. Hence, being a family is a skill. When you lose sight of the act and practice of becoming a family, members can be hurt.

When a family is hurt, members cry. When a family is in pain, members sigh. When a family is in anger, members fight. We learn from the facial expression of family members, how they feel or fail to feel for each other. Not being able to recognize each other's feelings is a lack of awareness or a feel for other family members. This means to feel things is an event. To experience feelings is a goal. To feel contact, you need to aim the brain to move through a line of focus and grasp a sense of these feelings in the act to interact with them. You are doing this to learn how to deal with being hurt, which creates a state of pain, which may turn out to be anger. The learning cycle would look like this:

Human Systems Model 17

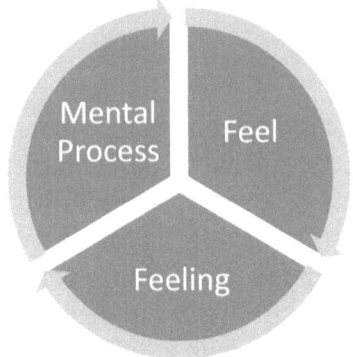

1. The event is to feel for
2. The goal is to accept feelings for
3. The affect is on the mental process variables between the event and the goal

As you feel the contact, these feelings are aroused. You move through them with the intent to learn how, and the experience affects how you come to think through thought. Hence, the way you become mental is through accepting how the contact feels, long enough, to process the interaction as feelings. The brain then moves through the act of becoming thoughtful, which is to be mindful of these feelings and their possible affect on behavior. Your anger may turn to a smile, your hurt may turn to a laugh, and your pain may turn to a shaking of the head as you act through feelings of thought. Hence, acting through the feelings changes the behavior through signs of care.

You need a system, strategy and process to do this type of work. In each step, the physical state of your family is being exposed. However, when you apply the systems, strategies, and processes we have described the physical state of your family will show signs discipline. Discipline is the physical act to show signs of a caring state to observe, listen, learn, help, and lead through to improve an external image. This is why we think that it is important to accept who you are, reflect on what you are becoming, relate why you are here to self, other people, and the environment, and judge the things you come in contact with by the way they lead you to feel.

Discipline comes from feeling things out through a line of focus. The body senses contact. You feel contact through the senses. You focus to feel the contact move through the senses. You build the capacity and potential for a state of discipline as the act to sense, feel, and focus releases a controlled reaction. You the family leader are signs of control. Family members, in response to you as a leader, reflect signs of discipline. You the family leader are the mental state of your family. You the family leader and your followers are the physical state of your family. Discipline is the applied use of focus from the outside to the inside.

Family Mental State

Few parents may talk about being mental, but it is the one thing a parent must learn how to do the most, be physical, but mental. A parent has to be seen as a thinker, which means a parent has to be seen as a leader with a certain mental state. A parent's mental state is what sets a family up to live in a home, and to learn how to learn in a school. Most of all, a parent has to be a feeling person. Parents have to learn from the experience of contact, which actively engages the act to feel events with a state of mind to think through the experience, to respond.

Being able to feel things out, leads to a line of focus, which leads to thoughts about the experience, which is mental. For instance, a parent's hands make contact with a book. A sense of how the book feels is formed. The parent opens the book to look inside. The physical act to look inside the book shows intent. Thought is involved in the act to feel the book with the aim to think through these feelings to learn. In other words, the sense, feel and focus cycle creates an aim that the parent uses to think, which means to become mental.

In this sense, the parent becomes mental in the act to control how it feels to experience holding the book. When the parent opens the book to learn what is in it, these are signs of discipline. Discipline is the physical side of the act to control the mental state inside the body. Hence, to be mental means to display a physical state of discipline. For instance, the parent's eyes are aimed in the act to read the book. The child can feel the parent's brain, through their eyes being aimed; and the parent's body, through their hands being aimed. The parent has shared a physical and mental state with the child.

Few people can talk about family-centered learning in this way, because we rarely talk about how a parent's act to become mental has to be modeled for the child to feel. We are not asking the child to think before they gain a sense of feel for the parent's role. We are asking the child to feel the parent's feelings for the role, and to use their response to send feelings forward that transcend the act to think. Hence, the child can feel their brain and body becoming more aimed, in response to the parent's modeling. How the parent and child relate to each other, as leaders and learners is important.

A parent's work with children has to center the act to learn how to become a learning family. This is why it makes sense to practice showing signs of care as a human act, to create meaning. The parent gains knowledge from learning, to learn as a result of interaction between the learning and the children. Hence, the children are learning how to make contact to interact as a learning family of experiences that connect them. Family-centered learning is the connection between the parent and other members to function as an interrelated and interactive whole to learn how to improve order and decision making, through the process of energy, action, and feelings.

The parent's energy leads the family to feel, as input. The parent's input is the experience of feelings for the family to experience as throughput, which is a process cycle of environmental awareness. Hence, the act to feel moves as energy and feelings from the parent, which is intended to cause action, as output. This means the event is environmental, the goal is learning, and the affect is awareness. This forms the boundaries of the action learning continuum we use to help the parent learn how to regulate contact to influences good behavior through the systemic structure of acting.

The parent assumes the leadership role, because it entails the cause to teach the child how to live with them as a learner. Hence, family-centered learning is how the parent learns to lead a child to learn, how to work with them. The act to live is aligned with the act to learn as a school of thought, learning how to live and learn as a family. This is a family-centered learning plan, to help a child come to think and respond to a parent's values. For instance, the name given to a child is relative to the parent's state of mind.

A parent names a child based on the way they feel. Hence, a child's name is intended to send thought backward to help the parent deal with their feelings. The child's body is a sign of the physical state, and the child's name is a sign of the mental state. This is the way the child is felt by other people. The value of the child to the parent's sense of family is expressed through the name. The way the parent, child, and other people feel about the name leads them to think with the parent's intent in mind. A child's name is intended to structure thought.

The name given to a child allows a parent and others to assess the needs of the child. Such as who they are, what they are becoming, how they are feeling things, why they may feel the way they do, when are they at ease, and where they think they belong. The way a parent sets up a home to be experienced by family members will help the child learn how to live with them, and will help the child learn how to learn from them. In the role of a model, the child is learning how to take commands from the parent. Hence, the child's name forms a memory structure for the use of words and talk.

Human Systems Model 18

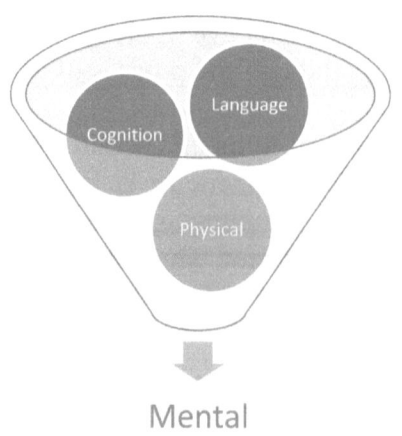

The memory structure supports the way a family reflects on the use of language, cognition, and being physical as experiences in the process to become more mental.

Family-centered learning is an educational method designed to lead the child and parent to form their bonds around the art and science of living to learn, learning to think, and thinking to respond, and responding to become more mental than physical. The study of the words *language*, *cognition*, *physical*, and *mental* leads the parent to become a participant in the assessment of how these words relate to setting a family up for success. In this sense, they are tools to assess how the parent and child lives, learns, thinks, and responds in their home, school, neighborhood, and workplace networks.

The parent needs a school of thought, like family-centered learning, to become more educated about how to set a family up to be governed. At the formative level, these words set up values for their use in everyday life: *language*, *cognition*, *physical*, and *mental participant assessments*. To the parent interested in learning how to live through a practice, these terms may take on business meanings. For instance, it is in the parent's interests to integrate practices into the role of parenting that will help them monitor their goals for self, other family members and helpers.

A certain tone has to be set. The parent models a behavior that helps people learn with them, as the leader of the family business. The meaning of the parent's role is drawn from a family-centered learning plan which is a vision on why they are here. Some aims may be to learn how to use talk, to learn how to feel things, to learn how to show signs of discipline, to learn how to show signs of control, and to learn how to assess thought. The parent's skill sets have to be set up to balance their aim for successful family, education, and leadership and business outcomes.

Family Social State

Most of the problems your family will face are in the home, school, neighborhood, and workplace network. A family social state refers to these places where you must live to learn how to think and respond through your constant interaction with them. The words "live, learn, think, and respond" are easy terms to keep in mind. You live in a home, you learn in a school, you think in a neighborhood, and respond in a workplace. This is a system, strategy and process, and, they all relate to social thought.

Your family can use them in each of these places to talk, bring feelings to mind, to build on a sense of self, or to think things through. The disconnectedness of them as strategic social places where wealth and status are gained, make them real aims to plan how you want your family to be set up, as a learning system. The social value of the system is marked by the order of need: you need a home you can live in, you need a school you can learn in, you need a neighborhood you think in, and you need a workplace you can respond in.

The social state of your family is set up by the way you choose to structure the ways in which they will make contact and interact with these places. The words come easily to your mind: events, goals, and affects you can plan to move through with a language, cognitive, physical, mental and social strategy for thought. Once you set up your continuum to make sense of time, space, and energy their systemic contact and interaction will improve your capacity and potential to move through human networks more efficiently.

You will feel and experience a synthesis in physical, mental, and social thought. Your family will be aimed more ready to compete and improve how they live, learn, think, and respond alone and together as they move through these circumstances. Hence, creative and improved methods are easily being recorded and structured to allow feelings and thought to take place. The connection of the family social state from one member to the next is bridged through the use of talk and memory from real life experiences. You talk through the experience of parenthood to help your child keep in mind the words that connect them, which cause them to feel and relate back to you.

This is the act. To help your family learn how to cooperate with the people in their home, school, neighborhood, and workplace networks, they must feel their need to compete through their capacity and potential to structure feelings of thought. Family social thought advances when ideas work out in connection with feelings of care that are used to think through the people. Hence, you sense, feel and focus how you and your family will experience each other, and people. This means you work from the act to live to learn how to think and respond to contact.

Family members cease to move through conflict haphazardly trying to think before they feel things out to learn. If you try to live without discipline and self control, anger may kill the growth and wealth in your family. Especially in a family that fails to realize how the network works to get rid of people that do not know how to interact. Your family's interaction is physical, which means it has to be more mental because the contact is social. All this movement between time, space, and energy is natural and predictable. You set your family up to feel things, but to talk through any signs of a behavior that will affect its social standing.

This connects their acts to live with direct structures to the use of language, cognition, physical, mental, and social practices to feel things through and to think things through to learn. Real feelings are in mind. Real thoughts are sent forward. You can assess feelings through an order to do something. You can assess thought through the study of a response. If you can show signs of care, they can show signs of care.

You and your family will need to work as a learning team to bring these values to life in your network. The following lists of words relate to the family social state:

1. Status
2. Modeling
3. Relationships
4. Wealth Building
5. Leadership
6. Followership
7. Contact
8. Interaction
9. Cooperation
10. Quality of Life

These are social values that influence how you move through home, school, neighborhood, and workplace networks to live, learn, think, and respond to self, other people and their environments. To be realized, these values must be balanced by streams of systemic action. In each word above, we use the word *feel* to gain a sense of them.

You need to feel status, you need to feel modeling, you need to feel relationships, you need to feel wealth building, you need to feel leadership, you need to feel followership, you need to feel contact, you need to feel interaction, you need to feel cooperation, and you need to feel quality of life to move these feelings to thoughts about their benefits and consequences. This new era of family-centered learning in which we live is one marked by how well we come to thought through the processing of our feelings.

For instance, people ask, can you feel me?—not can you think of me? The reason for this order is that the brain receives feelings, not thoughts. Thoughts are what grow out of the mental aspects of moving feelings from contact, to interaction, to cooperation, and to participation in the act to think. If you ask your child to go to the store, they feel your words for meaning. You will look at their body to gain a sense of how they feel, not think. If you want to learn how your child thinks, you will look for thought in their response.

Family Environmental State

A family environmental state is created from you and your family's capacity and potential to use our systems approach to improve signs of social control. When your brain, body, and senses are organized to feel and focus these feelings, we call this effect systems feeling. Family environmental state refers to you and other family members and the social places we use to practice learning how to live, learn, think, and respond to tension, stress, and pressure. As we have explained, self (you), family members, and the home, school, neighborhood, and workplace network are social environments that cause feelings to be aroused.

You have to be a feeling system to move through emotion as a mental process. To have social control means you are moving through self, other people and their environments in a state of focus that help you, to reflect and relate, to the consequences. Your internal environment is now set up for systems feeling, which means to feel things with intent to bring about a past, present, and future sense of an event to learn from the experience of becoming mental in the process of thought. As you learn, as an adult, you are a systems feeling person. As you learn how to lead your family (children) to learn how to move through their feelings, they are feeling systems.

The differences between an adult and a child are clear. As an adult, you have to know how to move through your feelings. As a child, youth or young adult you need to learn how to move through your feelings. In either case, you have to reach this point. To be a sign of emotional control, the internal state of your sense, feel, and focus cycle has to be working in harmony with your receive, process, and respond cycle. To live with self, family, and other people free of emotional breakdowns, you have to learn how to move through these process cycles at home, at school, in your neighborhood, and in your workplace environments, for social signs of internal control to become structured.

From this perspective, those initial cognitive experiences have been set up for change through these practices. You have built up new cognitive experiences through your act to practice living each day to learn from experience how to become more informed about these systems, strategies, and processes. So again, when you reach this point the consequences of not sending your feelings forward to receive thoughts backward is registered as a reflective and relative state of mind. This means you reflect on the consequences of not allowing your brain, body, and senses to practice working through the problem of feeling things that may cause you to feel upset.

Human Systems Model 19

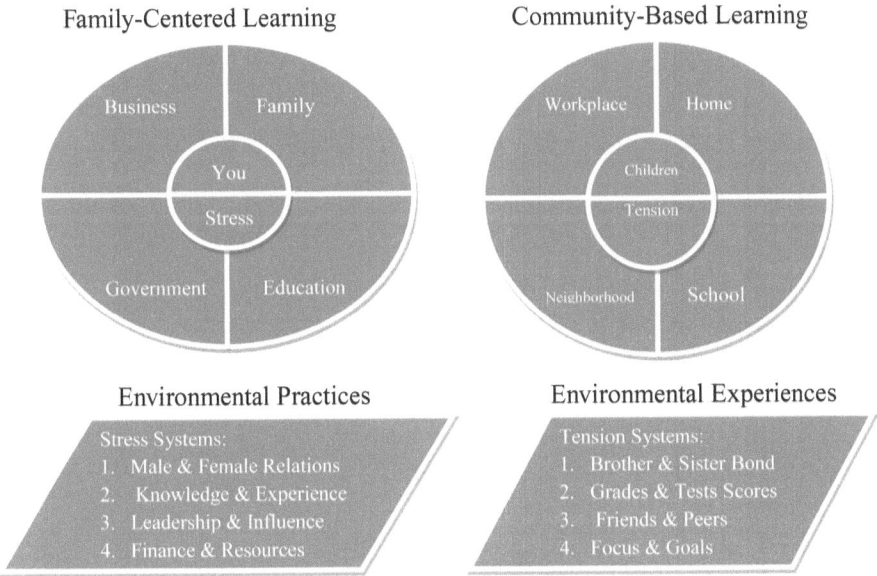

Family-centered learning is an environmental learning strategy designed to help you focus how you grow, mature, and develop through our systems and processes. This is shown in Human Systems Model 19, above. You need to build the capacity and potential to process information from each of these fields. Hence, the strategy is to offer a structured way to learn from the things you will have to do naturally to live, learn, think, and respond to stress.

On the other hand, community-based learning is environmental strategies to help you provide a line of focus for your children to grow, mature, and develop though our systems and processes. This is shown in Human Systems Model 19, above. You will need to help them build up the capacity and potential to process information on how to live with you, learn with you, think with you, and respond with you in relation to home, school, neighborhood, and workplace experiences. Hence, this strategy offers a social and economic structure for moving your children through the things they will naturally experience in your family system as tension.

Human Systems Model 20

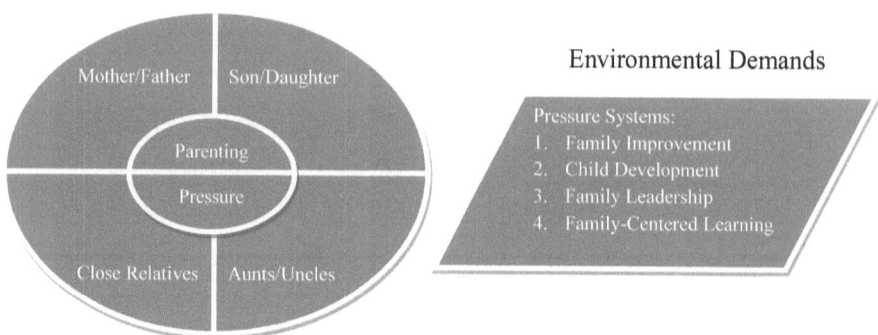

Family development is the environmental learning strategy we set up to help you talk through a science-based approach to reduce family decline, school failure, delinquency or crime, and unemployment or poverty. You will need to build your contact and interaction with self and other family members to help them become assets in your strategy to create safe places to practice living, learning, thinking, and responding to the roles and functions in parenting. This is shown in Human Systems Model 20, above. Hence, this strategy offers change by learning how to talk through goal oriented practices that lead to changes in behavior. When you make demands on other people to live up to certain standards, as a parent you represent signs of pressure. This is a future point of view.

Curt Lewin (1951) described tension systems as fields of invisible space between the self and the external environments. We are using this theory to describe how children living inside your family system are signs of a physical and mental tension system. In theory, they are living inside your family, living inside of you. When you think of stress, we define it as an external field of invisible space and physics. You can feel family members and they are physical signs of progression or regression. In other words stress is physical to the senses as you extract data from outside sources. For instance, when you feel and can sense, a lack of focus from the people, they are higher signs of stress entering your system, which creates higher states of tension or emotion.

When we describes anger, hurt, and pain these may be signs of emotion being released through your tension system. As an environment within other environments, tension is internal and stress is external to the flow of emotion and sense data. However, as we have stated, sense data has emotion in it. Both ways, send and receive paths are either open or closed to the processing of tension and stress through the practice of participating. At this point, the learning cycle is complete. You know if you want to live through the crisis, you have to learn how to set up events that help you and your team practice participating with self, other people and their environments.

Family Emotional State

Family emotional states affect the way feelings flow through the team based on whether the sense, feel, and focus cycle is open or closed to the way the contact feels. When your family members are focused from the physical to the mental states of the system they are set up to enter and exit social places with a high sense of control. Social control then, is the result of focused states of discipline and self-control. Meaning, your family lives, learns, thinks, and responds through family values that govern their displays of behavior. This is not programming, because your family is learning how to live with changes being implemented as they move from one event to the next.

At this point in the process, your family has learned how to adjust to living with an attitude that makes them feel different, because of the way anger, pain, or hurt feels when you are emotional. You realize that anger is a sign of being hurt, and when you add emotion to the mix, you may be in pain as well. These are the things you have learned to do. You use talk, to talk to the brain and to manage body movements. You stage the experience of helping your family to learn how to help recall things to mind. You model signs of discipline to set up internal control. A focused brain and body is a sign of total control over the senses. You practice using these skill sets in your home, school, neighborhood, and workplace networks.

This is important because in every event, your family is being tested by the emotion that exists in the contact itself, and their subsequent interactions with self, other people, and their environments. Hence, it makes sense to learn how to practice dealing with emotion at each phase of the system: language, cognition, physical, mental, social, and now emotional control. This means that by this stage your family ought to have the capacity and potential to work through internal problems and external problems with a greater amount of control over their brain, body, and senses as a team.

When you look back on the words hate and love, you can feel the gap between them. You know to show signs of care you have to feel things out. When you resist the need to feel things, you show less signs of care. Care is an emotion that ties your feelings to thought. You do not have to love something, when you show signs of care. However, you do want to allow your senses to move how you feel to your brain where those feelings are processed. In other words, at this point the goal is to trust your feelings. If you trust your feelings then everything you experience is fed forward to your brain for processing. When you trust your brain, your brain by design, sets up internal defense mechanisms to protect and defend your body and your senses.

This is why doing this work is vital. As you become more mental your brain is using reflective practices to assess the consequences of participating in any event you or your family experiences. If you were not open to the experience of talk, then the new cognitive experiences would not have allowed you to move to the next step in the process. To move feelings forward, you have to be open to changes in your state of mind. This is where many people fail to recover. The need to resist is greater than their need to accept, because bad experiences are red flags to the mental process cycle.

You reflect upon contact, which also means to relate any events that may be connected with the experience. When you reflect and relate, you are judging, which in this case means to assess. If you choose to comprehend your need to feel these feelings, then you receive, process, and respond to change. When you are hurt you resist because of fear. When you are in pain you resist because of fear. When you live through a state of anger, you resist because of fear. You are more apt to resist, because you are in a state of mind that makes you think as though there are sufficient levels of control.

The truth is, however, you resist, because of the threat of losing control. In other words, you resist because you do not want things to change how you think to feel things, from the outside to the inside of your body. You are more willing to live in a state of crisis, because you think you are protecting your sense of self. The truth is, however, that you are not sensing, feeling or focusing your brain or body. You are ignoring how the contact feels, to resist feeling things you cannot control. This is why signs of discipline have been taught as a physical skill. You cannot afford to ignore an event. If you do, then you are refusing to allow your brain to practice responding to the things you experience.

When you look at the brains capacity and potential to show signs of discipline, you feel change. The experience of becoming more mental than physical creates changes through the way you receive, process, and respond to tension, stress, or pressure. These are emotions in the body, in the family, and in their environments. When you choose to move through them, you are choosing to comprehend the need to make adjustments as feelings move through you, your family or their environments. When you look at the body's capacity and potential to show signs of control, it means you can feel the flow of energy changing how these feelings are being allowed to affect your state of focused thought.

You want your family to trust your brain. The way you move through emotion, allows them to learn by example. You do this, by moving through your feelings to store signs of love in your family system. Every member has to feel your love. Without this sense of love, then there is less reasons to reach for the words *joy, pleasure, excitement,* and *happiness*. You do not want them to move between states of anger and fear, pain and disappointment, hurt and sadness, and most of all, you do not want them to hate people. Without focused feelings of thought, the brain will cut off the flow of energy.

Your brain has to learn how to live with your body. Hence, you have to learn how to live with your family. Your family is the body; you are the brain of your family. Sensing when you feel agitated is a basic sign of care to move these feeling toward the brain, but through mental process cycles. When you look back to observe how your body responds, you are listening carefully to learn how your family follows your lead. Your brainwork is being interpreted, if you and your family have learned how to use these words as a strategy to accept each other's help.

Hence, you can see less resistance to your contact, and more control over their sense of feel for you. A family emotional state has to be set up, for this reason. Your capacity and potential to assert a sense of self changes, and those images come to you mainly from the values you hold for family. You will respond assertively in the above situations, only when you see yourself as a worthwhile human being" (Hopson and Hopson 1992). A family leader has to be easy to follow. We mean, easy to feel for signs of care. As human beings, we naturally feel for the people we care about. You protect and defend these feelings each time you work them through to a complete sign of care.

Without you allowing your body to be a sign of care, there can be few changes made to this end. A family leader has to search for ways to show new signs of care with each step or gain. This is how you change the things you can, to learn how to reach for your capacity and potential to live through your family system. Leaders cannot display anger as a message of control, because anger can become a fixed state of emotion. Leaders have to counter this threat with displays of joy to balance how these feelings are being transmitted. To be a family leader requires a special kind of feeling system.

Human System (Learning) Model 21

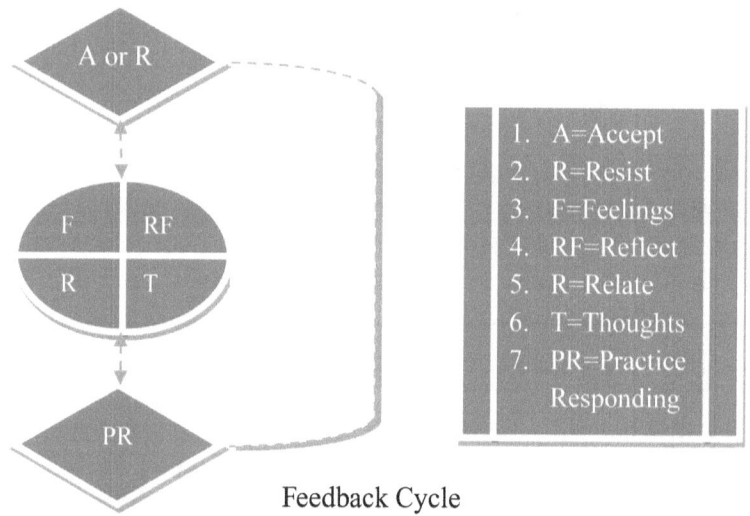

Feedback Cycle

The human system learning model above is comprised of seven elements, which work to form an orderly and interrelated model for how we may handle the whole list of functions (Slaton 2002). The feedback cycle is external to the internal functions of the model. Any event is responded to from the inside to the outside of the self through the sense, feel, and focus cycle, and creates a feeling system more open to the threat of change. The feedback cycle is intended to store the experience of the act to practice responding to the circumstances surrounding the event. The practice of Progressive Investing will continue to be explained throughout this book.

Family and Sensory Organization

Whatever family you are in, you want to be loved and you want to be liked. Most family leaders are interested in organizing their senses around feeling things and thinking things through, so they want to be seen and heard in an effort to love and be liked. Similarly, most parents want to learn how to lead a family, to live together, feel and think about each other in their own way in hopes of structuring how they love and like each other. When you are in crisis, there are things you do not want to see or hear, because of how they may affect the way you feel and think.

You have eyes, but you do not want to use them. How can you have eyes, but not want to use them? Eyes are used to feel things. In this sense, to have eyes does not mean you want to use them to see things. Eyes are in the vein of sight. To see with the eyes is physical, and to hear with the ears is mental. But you can have eyes and ears, and not want to feel things. You may see me walking your way, and hope I do not see you first. You may hear my voice and look to elude me first. When you do this, you are trying to hide. How does it feel when you hide?

A family has to make sense of how it feels to be close. But then, as a family grows in age, it changes with age and circumstance. My brother called me to tell me "I love you, but I like you as well." You know he said to love you is not the same as to like you. I can love you, but not like you. What I like about you is how you carry yourself. The way you try to make other people feel when they are around you. We have to model a sense of self that can be learned in terms of time. He values the time I have used, to stand as I am.

For instance, I am not afraid to recall how living in our family made me feel or caused me to think. Even if I am hurt by these feelings, they are not allowed to influence how I think with other family members. In this view, my sense of care for self and others rises above pained thoughts. But this means, I had to come to terms with the person I am. In doing so, I had to come to reflect on these experiences as points in life where I began to live and learn for more than just myself. I had to see my sense of self in relation to other people and their ability to feel me, and then think about how I make them feel.

As your family changes, you must feel these changes to learn how to think them through. To make sense of the changes in you and other family members, the way you practice feeling things and thinking things through must be organized. In other words, other people have to be able to recognize the things you do across a continuum of events to make sense of the person you are becoming. These are signs of leadership.

My ideas on family leadership dates back to my childhood, I can easily recall studying my mother and father's responses. Leadership is thought to be a vital process of reflection and action through the works of Plato, Caesar, and Plutarch (Marzano, Waters, and McNulty 2005). Hence, leadership in a family is no different from acts of leadership in education. Leadership is thought to be the cornerstone to successful functioning (p. 5). If we consider the perceived importance of leadership, the parent, the nun, and the teacher emerge as essential figures in the growth and practice in working with children as their leaders.

This dates back to antiquity. Leadership started in the home, moved to the neighborhood, and now reflects its links to the school. If you believe that a parent is not a child's first and foremost leader, then what role does the affect and effect of time play in your current judgment. To what extent does the role of the church reflect on this affect or effect on the family learning system? In fact, the separation of church and state reflects the takeover of the child's learning system by school leaders.

As a child, you search for ways to help your parents understand the things you do to gain their affection. What is important to your parents become important to you? Leadership, or who was in the role to lead, was very important to my parents. That is, where we were headed needed to be clearly stated, in terms of how we were going to get there. I was hurt by my mother and father's breakup. Given the importance of their modeling, it is no wonder that I think an effective parent is one that shows signs of care for the child's brain through the use of talk to lead.

I had to learn how to love me, to live through feelings of pain. I love where I am, but I do not like how I got here. This is because when you grow up hurt, you hurt other people to make you feel a sense of pleasure from their pain. If you feel things like me, what my brother was thinking was he likes the way I live through signs of pain today. But I like the way I learned how to lead, from my mother and my father. And hence, if I had not become more organized from the inside to the outside, I may not have moved through the pain to dump those feeling of hurt. When you hold on to them and they reduce your ability to sense, feel and focus how you live with your sense of self.

I could look at my eyes and sense the hurt. I could feel the hurt, when I found it hard to hold my head up to see and hear. Sense data has to be received, processed, and responded to by the human system. Sensory organization then, is how you use data as a physical and mental purpose to respond through signs of care. You do not hide how you feel; you act through the way you feel. You feel the hurt, but you respond to live through it. Sensory organization starts with how you react to feelings of hurt.

To us as family leaders, the phrase *growing up* means to live, learn, think, and respond through the hurt. Hence, family and sensory organization starts with how parents parent a child that has been hurt by major life events that affect how they receive sense data. Home, school, neighborhood, and workplace events lead to such feelings: I do not feel good about my family. Nobody in my family seems to like me. I cannot think when I am there. Why do they send me to school? Why can't I just be left alone? I want to be loved. I prefer to be loved than liked, or liked than loved.

Typically a parent has to organize how they move through these acts to live in a home, to learn in a school, to think in a neighborhood, and respond in a workplace in ways that connect their meanings to the child's experiences. Over these events the child learns how to make sense of the people they meet in these places. It is the people in these places that cause a child to feel for signs of love and to think through signs of being liked.

Family and Green Science

When you explore your home, school, neighborhood, and workplace network from the sense of how you feel in a family, the task to learn how to help improve the things you can take on the need to reprocess signs of care. This means you feel the need to try and make people feel a certain sense of uplift when they encounter you. Green science is a shift in terms of green technology to recognize that a family has to generate signs of care that provide evidence through safe exposures to people that will improve the way you practice learning how to love self, other people, and the world.

How a parent grows a child as a human system, changes the living and learning process. Parents use the sense, feel and focus cycle to set up the receive, process, and respond cycle for the child to grow into the family as a green technology. The parent and the child are set up to grow together as living and learning systems that use the home, school, neighborhood, and workplace to practice this new approach. The parent may not be a scientist, but the parent is a child developer, in this same sense.

Sense data has to be received, or the child has to learn how to receive sense data, to feel how the process to respond requires focus. The parent, as a child developer, studies the physical aspects of the child's behavior in an effort to understand the feelings being aroused through the act to sense, feel and focus which requires discipline. Hence, the parent looks for signs of discipline in the child's response to what they see or hear. To us, setting up the child's use of sense data builds on his or her mental capacity to move from feeling the things they see and hear to living and learning through the things they process as a result.

In this same sense, to live and to learn is not the same thing. To live means to sense, feel, and focus the act and to learn means to receive, process, and respond to the act. During these process cycles the parent and the child are becoming more informed about how to live alone and together with the intent to improve the things they can, as they learn ways to move through them. The knowledge that is acquired through the feel and process stages transform them as knowledge is processed; they may be called knowledge processors in this same sense.

This means thinking things through has occurred at the same time, to form the required connections to the experience of these acts to live and to learn. The act to think may or may not unfold the meaning to the process without the parent's signs of self-control and the child's signs of discipline. The parent's love for the child, the parent's thoughts of the child; and the child's like of the parent, the child's thoughts for the parent; move from feelings of self-control and stages of discipline in the formation of knowledge as ways of learning and recognizing their history.

In short, parenting has to be set up as an event with the goal to lead the child to sense and receive an affect. The parent not only has to feel the child's act, but think through the pattern that results from their exchanges of physical and mental contact to prepare the whole child, to live and to learn through them. How does this feel? What do you think? Why do we need to live and to learn? Who is the child? The parent must grow as the child learns how to live within the family to reprocess these responses in home, school, neighborhood, and workplace environments.

Feeling things through produces thinking, which changes the act to live and to learn from the experience of parenting. To us, this is green science. To grow as a parent or child with intent to become or to improve the flow of sense data to the brain provides the connections for the human system to emerge as an advanced response to living and learning from experience. The experience is defined by what is received from the flow of sense data to the brain, and requires focus to be felt.

1. Yes, the parent and child do need to know what it feels like to live through acts.
2. Yes, the parent and child do need to know how to think things through.
3. Yes, the parent and child do need to know why they live and learn as a team.

To love someone is not the same as to feel for someone, because to feel for someone requires self-control to learn from the act. To love someone without a sense of self-control, means there is a lack of discipline in the act to make sense of the person. These are strategic process variable in family relations—loving and liking, being mental and being physical. Feeling physical is the same as being mental in this case, because the parent and child have to recognize each other's identity through signs of self-control and discipline interchangeably. Hence, there is a problem when the brain and body are not set up to sense and feel things received and processed into states of awareness.

Feelings of love are less grounded in the act to live through the experience of thought, which include the study of signs of care about being liked in the physical sense of becoming and growing more informed. When you are hurt, the transfer of sense data to the brain is affected by the state of how you are living and behaving. When you are hurt, the act to receive sense data is affected by the state of how you are learning and problem solving. Love has to be felt, being liked has to be experienced. This means signs of self-control and discipline have to be captured through the experience and reprocessed as an event.

Moving forward, the aim or the event is designed to change the affect from feelings of hurt and anger to thoughts of being loved and liked. These are new cognitive structures, or new building blocks that allow the parent and child to move or change the way they feel and think with the brain, body and senses being reorganized to act as a human system. The parent and child's effort then is to aim talk at the brain, to focus the body and to use the senses to become and grow more self-controlled and disciplined.

Family and Education

Family and education always represent two distinct and separate schools of thought. The family, as a school of thought is dismissed as nonexistent or inadequate. Education as a formal school of thought is valued and promoted. This is a narrow formulation of a problem that threatens the development of the whole child and does not engender feelings of being accepted and valued that are necessary to prepare a child's brain, body, and senses for the experience of school. Hence, how a child learns and how a school wants a child to learn, will always reveal the differences that exist between these two cultures (Logan, Freeman, and McRoy 1990).

"In the ecosystems perspective, the child, the family, and the school are always in reciprocal interaction with each other; notwithstanding the tendency to keep the school somehow separate as a system that merely receives children to provide the service of education" (p. 124). Most children in crisis, as a group, learn to hate school as their school experiences reveal a poor acceptance for who they are, and less value for the type of family they come from. "The crisis of hatred is a developmental crisis for the group; it has all the recognizable characteristics of a virulent negative transference" (Agazarian 1997, 106).

Crisis is a term used by social workers in two ways: (1) as an internal experience of emotional change and distress and (2) as a social event in which a disastrous event disrupts some essential functions of existing social institutions (Barker 2003, 103). The term crisis generally evokes an image of any one of a number of very negative life events (Dattilio and Freeman 2000, 1). According to Segal and Yahraes (1978), "Inadequate education, poor physical environment, meager and disorganized life-styles, poor physical and mental health, large families, broken families and low income" (p. 261) are factors that move emotion in social events as potent and pervasive threats to a child's development.

In California, for instance, the percent of economically disadvantaged children in public schools has been steadily increasing (Taylor 2009). "Approximately 40 percent of public school children were economically disadvantaged, as measured by eligibility for the federal Free and Reduced Price meal program, which requires income to be at or below 185 percent of the federal poverty guidelines" (p. 4). Hence, "a crisis may, however, also relate to circumstances or experiences that threaten one's home, family, property, or sense of well-being" (Dattilio and Freeman 2000, 1).

This pertains to the child's sense of self, other people and the environment as a physical group of life threatening events.

When you explore your home, school, neighborhood, and workplace network from the sense of how you feel in a family, the task to learn how to help improve the things you can take on the need to reprocess signs of care. This means you feel the need to try and make people feel a certain sense of uplift when they encounter you. Green science is a shift in terms of green technology to recognize that a family has to generate signs of care that provide evidence through relevant exposures to people that will improve the way you practice caring.

To remove a family from crisis requires special learning to improve parent and child relations in connection to school and neighborhood affairs through which people outside the home can work with the family to share values, purposes, and objectives. Building a learning family requires enhanced connections among parents, teachers, and neighbors who live together in a neighborhood or work with the family through school to share events, goals, aims and values. When parents, teachers and helpers are involved the child's chances for success in school increase. In effect, a family-centered learning community is built for people to work together through shared values. One person leads and another person prepares to observe, listen, learn, help, and lead to broaden the capacity and potential for successful connections. Hence the parent and child learn how to learn with other people in meaningful areas of need (Garcia and Hasson 2004).

In a learning family, a child enjoys higher academic gains and has better behaviors and attitudes toward school, because you are constantly talking and working with their brain at home, but in connection to school. The education of the child becomes more of a process. You are seeking to address the needs of the whole family through the child's need to do well in school; family-centered learning recognizes the strong link between family members and children thus their education process (Garcia and Hasson 2004). Each participant works to form an intergenerational work group to reinforce this effort to live, learn, think and respond as a team.

Family-centered learning from this point of view is parenting by doing. Systems are set up as a plan to activate events and goals that can be assessed as curriculum elements. Parents and children learn how to look at family issues, problems, and concerns as a set of interrelated elements; a home, school and neighborhood need to be skillfully connected to produce the preferred outcomes, and with the input of all relative stakeholders. Team members relate to each other as human systems. In other words, every human being and every group of human beings make up a system (Kuhn 1975).

We have integrated language to observe cognitive experiences and to change physical behaviors as basic elements to the mental process that calls for learning how to live by doing what human systems do. Learning how to learn to be learning to learn methodically interrelates. People identify with the physical response of the act to learn, which is social. Who the parent and child is becoming, shapes and influences their use of schooling (Wortham 2004). Hence, learning changes not only what the learner does, but how the learner acts to become more aware of the person they are, and the person they are becoming.

To learn how to practice becoming mental, changes how the learner feels about learning. "Learning can change identity and self; learning transforms who the learner is and what the learner can do, it is the experience of identity" (p. 716). As social institutions or major life events in the lives of parents and children, schools are important, but only to the extent that they are part of a larger thoughtful pattern (Sizer 2004). According to Banks (2002), many schools continue to function on outdated curricula and structures based on the theory that only a small elite group of learners will show the drive to gain academic success (p. 37).

Do you talk to your child's brain, about low school achievement and the threat of dropping out? Poor school experiences, such as not being able to read and decode words produce resistance (Biancarosa 1995). Not being able to move words through the senses and comprehend their meaning is deflating to a child's drive to learn how to gain something from the experience. Reading, reflecting and writing are the basic building blocks of talk through the use of words. You talk to your child's brain to aim the senses, because learning is an ongoing process of feeling noises, sounds, signs and symbols to make sense of them.

Your goal is to prepare the child's brain to read things, to think about things, and to talk about things. According to Biancarso (1995), reading is thinking cued by written language (p. 24). How an adult chooses to respond to a child in crisis will affect their emotional, social, and cognitive development. Everything the child takes in through the senses arouses feelings. When a child is asked to focus to perform an act they cannot process, these feelings become social experiences. Cognitive structures are basically stored memories, and in this case, they are hurt by a lack of capacity to feel and process the act to focus sense data.

When a child in academic crisis frequently sense signs of disapproval, they are more likely to store harmful experiences and resist learning how to move through those circumstances again. Emotions that are not moved through the mental process can create deficits in a child's intellectual abilities, which hurt his or her capacity to learn because of the pressure. The organization of the brain, body and senses reflects its experience. If a child's experiences are characterized by fear, anxiety, stress and hopelessness, the chemical responses to these emotions become the most powerful architect of the brain (Scherer 2005, 24).

You talk to your child's brain to set up multiple paths to intelligently respond to tension, stress, and pressure. Learning how to use talk fits your child with a natural way to practice releasing tension. They sense, feel, and focus the brain and body to release feelings. This allows him or her to free up space to experience stress through an open path that is naturally designed to channel feelings of stress or pressure. The path is basically strengthened by its connection to a physical identity (you and your child's sense of self). You are mental; your child is learning how to become more mental than physical. Intellect is built and sustained through acts of discipline and self-control, which supports this flow of feelings.

We have just released our thoughts on the use of practical and analytical intelligence. The act to sense, feel, and focus is to set up paths for practical intelligence. The act to receive, process, and respond is to set up paths for analytical intelligence. Howard Gardener redefined the word *intelligence* as a biological potential to process information that can be activated in a cultural setting to solve problems or create products that are of value in a culture" (Reeves 2006, xiii). To change poor behavior, your child has to learn how to release feelings forward to move emotion through the body to build up a capacity and potential to strengthen signs of discipline and self-control long enough to feel the process of thought working in reverse.

Your child learns by doing things to help you lead them. The emotional self moves through the practical self. The practical sense of self grows through its connection to the analytical self. The analytical self feeds feelings forward to the senses. The senses are armed with instructions on how to feel the experience of contact through the practice of reflection. Physical awareness and mental processing reinforces the aim of the senses to feel, focus and release talk to build the desire to write and think through printed feelings. Learning how to do things to practice becoming mental prepares the analytical self to respond.

Your child is learning how to practice receiving sense data as input, transfer how it feels to the process of throughput and focus feelings to respond as output. In the next step, the brain is open, because the child is looking for backward feed. Things that can be directly observed through the senses, is feedback. This allows you to set up the child's capacity and potential to release data through talk, reading, writing, and acting to methodically respond to tension, stress and pressure. Recall that, learning how to comprehend sense data is an ongoing process of thinking things through. Your child's brain is a real and logical message system that sends and receives.

The school system needs to be reformed, because it is not keeping up with the times. Why pretend that our school leaders and politicians are trying to improve our schools? (Reeves 2006). Many teachers continue to work in isolation from other teachers, which breaks down the practice or working together with other elements of the school system or the larger thoughtful pattern. What happens to student learning and teacher confidence where there is a breakdown in the flow of public practices in and outside the classroom?

Labaree (1997) provided the context for this discussion on out dated curriculum and school practices. "The shift of behavioral measures to cognitive measures is still a psychologistic and formal nationalist model (p. 151). He asserts that alternative approaches have emerged, but are currently operating only at the margins to central structure of research on teaching. The current state of standards and teaching in our public schools resemble the "black box" model that scientists use to study unknown phenomena (O'Shea 2005). In this model, identifiable inputs and outputs of a system can be seen and described, but the process that gives rise to the outputs is the least understood part of the system (p. 19).

In educational research, positivism or scientific empiricism is described as the refusal to accept (concede) the status of reality to things not directly observable (Price 1994; Slaton 2002). "The most blatant expression of this view is behaviorism, which eschews mental and cognitive constructs (Price 1994, 64). There is talk about it. "What educators really need to do is come up with other, more useful and valid ways to provide evidence about their classroom competence. One straight forward way for teachers to secure evidence of their instructional effectiveness is to collect defensible data showing whether students have made substantial progress mastering significant cognitive skills" (Popham 2005, 80).

A child's cognitive ability affects identity, inspiration and interests (Leonardo 2004). You want to develop your child's cognitive ability to set them up for personal, academic, social, and occupational success. A family language system will not only frame the way your child experiences learning, it may improve their sense of self, intellect, character, and job readiness. Learning how to use talk create conditions for intellectual growth as a preparation process for critical academic language usage (Lippman 2008). The only way education can be a process, is when parent involvement is set up in connection with the teaching and learning process.

Your family-centered learning plan can change the pedagogical process from one of knowledge transmission to knowledge transformation. A quality-based educational experience is enhanced for all when you as the parent, is involved (p. 11).

Human Systems Model 22

For clarity, you and your child represent the self. As you learn how to use the system, your child learns how to become more informed through it. Hence, the Human Systems Model above explores cognitive experience through roles.

You the Parent/Self

We talk to you from a developmental perspective in connection to what a child, youth, or young adult needs to be able to do with their brain. They need to experience physical and mental skills by the time they reach puberty for identity and thought to work in unison as they move into young adulthood. This will allow him or her to grow up with a sense of purpose to focus critical thinking and reasoning skills to help create ways to help. In these early stages of you and your child's learning, you are the spiritual leader of your child's sense of purpose.

You do not have to be a psychologist to learn how to work with your child's brain. All you need is the desire, to learn how to help the helper help (Slaton 2009). If so, then you will live each day to learn how to become more informed as a learner and critical thinker to be diligent in your efforts to perform as a family leader. This will organize your senses of self as an event within the goal to be a sign of social and emotional skills to really affect you, as a careful thinker. Hence people will view your acts to live, learn, think, and respond to tension, stress or pressure as responsible, respectful, and right aims.

You need to want to be as good of a model, as you can be. You are your child, and families mentor. You need to show signs of discipline, because you know you are physical, and you need to help aim signs of control, because you know you are mental. This puts you in the pursuit of a healthy lifestyle for self, your child, and your family. To learn how to live as a family is such a strategic endeavor. This is why talking skills needs to be persuasive (Lipman 2008).

A sense of self and living right or wrong is related to you as a good or bad spirit. Hence, as a parent, your sense of purpose will play a key role in how you grow as a helper. Becoming a more skilled helper involves critical thought and acts of reason to create complex connections to roles in life that helps followers learn how to pursue states of joy, pleasure, and excitement. The goal is to act happy. Thinking about your own thought processes is metacognition, where reasoning evolves from being both inductive and deductive (p. 14).

Personal

A child needs a healthy relationship with family to enhance how they live in, learn in, think in, and respond in social places. Their transition to adulthood will involve four domains of personal experiences:

1. Living in a home
2. Learning in a school
3. Thinking in a neighborhood
4. Responding in a workplace

How a child develops physically occurs in connection to life inside a home. How a child develops intellectually occurs in connection to learning inside a school. How a child develops socially occurs in connection to thinking inside a neighborhood. How a child develops psychologically and emotionally occurs in connection to responding inside a workplace. Having good feelings toward life begins inside the home and having good feelings toward life, enhances the functional use of language inside the school.

According to Lipman (2008), this develops good emotions through a sense of well-being and good mental health. A higher sense of self evolves from: (1) learning strategies, (2) motivational control, (3) emotional regulation, (4) attention control, and, (5) cognitive competency (p. 9). This means, as a parent, you are a human developer (Slaton 2009). The goals you set are to influence your child and families behaviors. You want good mental responses; you need good mental responses from your team.

You as a parent are responsible for developing a lifelong learning structure to plan goals. As you are learning how to stage these events, you, yourself, must be living in the pursuit of a rich inner life. As you learn how to move the use of talk, to develop cognitive experiences, to grasp human physics, and to feel mental, deep and purposive connections are being made between your sense of self, other people, and their environments. These are all signs of care.

Academic

The case is of the child (Dewey 1990, 209). Your child has to feel things, to learn how to care. "The consequence of such a method is that while a pretence is made to reconstruct educational aims and values in terms of a developing society and of the individual and his needs, it is the values inherent in different formulated "subject" matters" that are of predominant influence in the formulation of the actual aims" (Taba 1932, 211). The child's brain, body and senses need to be prepared for contact and interaction with the school, because the school is not set up to prepare the child for the capacities to be exercised. The child's sense of self will grow at odds with their personal and academic senses of self, because the school is set for traditional cultures.

"How, then, stands the case of the Child vs. Curriculum" (Dewey 1990, 208)? "Unfortunately the tendency to separate these two inseparable agents in education is so deeply rooted that the problem has to be viewed as a conflict between the needs and interests of the child and the demands of the objective social and cultural values" (Taba 1932, 212). One school of thought is that the child does not need to know how to think, just what things to think. Our school of thought is that the parent and the child need to learn how the brain comes to think, to feel and focus the senses for the cognitive experience of life.

"It is very important to note that a genuine insight into the values and methods of thought as well as into the factual material of the bodies to tested thought, and an understanding of the attitudes, ideas, generalizations and beliefs governing social life today, as essential factors for the successful conduct of education" (Taba 1932, 213). "An interest in the formal apprehension of symbols and in their memorized reproduction become in many pupils a substitute for the original and vital interest in reality; and all because, the subject-matter of the course of study being out of relation to the concrete mind of the individual, some substitute bond to hold it in some kind of working relation to the mind must be discovered and elaborated" (Dewey 1990, 206).

Albeit, Both John Dewey and Hilda Taba presented these findings many years ago, they still are recognized as a strong effort to combine the human, cognitive, and behavioral domains through a similar analysis of academics. Their analysis of the value of education as it is currently being practiced by and large relates to the child's mental health and well-being because the business of educating children is to program the way they think. Dewey (1991) wrote, "Apart for the development of scientific method, influences depend upon habits that have been built up under the influence of a number of particular experiences not themselves arranged for logical purposes" (p. 144).

"Unfortunately, the history of science shows us that such methodological considerations in delimiting the units for different sciences have not been followed" (Taba 1932, 11). We have our sights set on learning how to help children that have been hurt learn how to experience their home, school, neighborhood, and workplace as a network. Thought about how parents and children that have been hurt think, is the reason we are here. All disciplines, intentionally or unintentionally, are centered around certain elements, certain specific functions and aspects (p. 10). "The technique of scientific inquiry thus consists in various processes that tend to exclude over-hasty 'reading in' of meanings; devices that aim to give a purely 'objective' unbiased rendering of the data to be interpreted" (Dewey 1991, 87).

How, then, are we to move toward an open dialogue between us and public school officials, when they do not believe the science they practice is backward? Even though, as Benjamin Blooms (1956) work now clearly shows, cognitive experiences produce behavioral outcomes that move the child the farthest away from acting and behaving as good humans do. Taken literally, such advice is now being carried out at the margins of our public school system's gates, but still not at the levels that are required to move the child's interests to forward feed, which is to treat the child with an education designed by humans from the experience of observing how humans live, learn, think, and respond as advanced living organism or human systems. Green science then, is how we build human assets as processors.

Social

Human systems research is the study of self, other people, and the environment (Slaton 2009). A high sense of self comes from learning how to feel things: contact, interaction, and cooperation. Your child needs learning strategies that seek to motivate them to learn with other people, in social places. The social realm is where data is captured by the senses and transferred between complex send and receive paths that are either open or closed to interaction. These are cognitive experiences that lead to the release of a certain pattern of behavior. It is the social pattern you want to look at for signs of how emotion affects the contact and interaction.

Your child ought to be able to communicate, which means you can process their response through the study of how they use of talk, writing, or acting. You look for levels of focus, formed from what the activity reflects. Your backward feed is social. Hence, you ought to be able to sense and feel multiple levels of cognitive competency or thoughts that are becoming more deliberate and controlled. This means the strategy is causing sense data to be reprocessed through feeling, reflecting, relating, and thoughts.

These are signs of character. The behavior reveals a social pattern of values that reflect and relate to thought that has been processed through the strategy to sense, feel, and focus feelings. In the social realm of human virtues, you look for signs of honesty, loyalty, reliability, trust and ethics. These are signs of how well you have learned to inform—talk to, read to, write to and act through your child's brain. Character building is a critical social practice of delivering a sense of purpose to your child's brain that helps him or her, to better control their body.

Keywords for your practice: *honesty, respect, discipline, self-control, responsibility, fairness, perseverance,* and *kindness toward self, other people, and their environments* (Prestwich 2004; Slaton 2002). Kohlberg (1981) transformed the knowledge and function of moral development in education, as though he were talking to the child's brain. To be talkative, cognitive, physical, and mental affects the child's identity and will motivate their interests to become more informed through creative language that reflect how they are living on the inside of their body.

You may be formatted to talk at your child's body, because their body holds social values to you. We may have all been taught to talk at the child's body, because behaviorist that built the system of education did not believe or value the child's brain or the need to aim lesson plans to reach it. This means that you have become more involved to encourage educators to consider your child's feelings in connection with the way they are asking his or her brain to be used. They have to listen to your child, to learn how to teach to his or her brain, not to their body through the use of the senses.

Many children are in academic crisis today. Hence, they have built up a resistance to classroom learning, classroom motivation techniques, classroom emotional controls, classroom attention controls, and the most vital, classroom cognitive feelings. Following more than fifteen years of research with children and families in crisis, Slaton (2009) wrote that teachers have to learn how to talk to the child's brain. In other words, the teacher has to learn how to plan lessons that aim to reach the child's brain for responses to their pressure. These things take place in the social sense.

Signs of character from your child's body are what the teacher looks for. The problem is that a child's sense of character has to be built up. The behavior that a child releases is shaped by how people cause them to feel and respond to them. If the child does not feel respected and valued, it changes how their feelings are to be released. Hence, the behavior patterns that become more social have less thought in them, since the child resist to protect their sense of identity. This all points back to the growth of mental confusion Taba (1932) wrote so strongly against. Being a parent is a lot of work.

Occupational

What is your work? Leonardo (2004) wrote about education as a process that could not exist without the parent's involvement. He is writing from a critical social theory perspective on how to change the pedagogical process from one of knowledge transmission, to one of knowledge transformation (p. 11). In this sense, the transmission or information to acquire school knowledge is related to oppression, and the transformation of information to acquire school knowledge is related to emancipation. In this view, education must qualify as the experience of knowledge.

To oppress means to subjugate the process and to emancipate means to release the process to experiences of knowledge for the workplace. You have to combat this premise with the knowledge you are learning about why the home has to be a parent's workplace. Knowledge or know-how is connected to certain levels of intellect that strongly influence learning in a family (Ackerman 2003). The goal of your workplace is to move your family through academic, occupational, and vocational context through the use of your home, school and neighborhood as a network.

You build up strength through the use of your home as "the" workplace, is the first goal. The word work means to do something that will require learning from the experience of knowledge. Lippman (2004) wrote that work-based learning experiences build up practical knowledge of the job. You have to sense, feel, and focus these feelings to connect the home as a workplace, and to see it as a learning environment for your family to experience. The goal of the network is to integrate home, school, neighborhood, and workplace experiences to better inform and prepare the self, the child and the family for work.

To learn how to live in this way, learning changes not only what the learner knows, but also "who" the learner is becoming (Wortham 2004). "To learn is to take up a new practice, to change one's status in a community" (p. 716). Learning then, is an experience of knowledge that can be released as learning from a state of identity through a state of action or performance. I am on the inside; the parent works to identify the meaning and cause of certain feelings of thought. Which means the parent changes who they are and what they are becoming based on the experience of doing this work to learn.

The parent is capable of using situational cognitive experiences to learn with the people in their home, school, neighborhood, and workplace network when they can show these skills. This reflects that the parent's brain is engaged in the practice to learn how to set up and move through academic, occupational and vocational context. Learning how to fuse personal, academic, social, and occupational practices with the home, school, neighborhood and workplace network shapes the role of work for the parent.

This means that the feelings of children are moved toward both the social and academic context and the thoughts of parents are focused toward the personal and occupational context, to truly take advantage of their need to grow, mature and develop as a team. Having this capacity to experience knowledge forward, and the potential to grow as a learner and grow as a leader, builds the right attitude and is a skill. Hence, it is a skill to know how to learn, and it is a skill to know how lead. Together, how you learn and lead affects the way a child and parent experiences their home, school, neighborhood, and workplace network for personal, academic, social, and occupational knowledge.

The process, the practice or the method of doing things to learn, are the keystones of your work. People will learn how to work with you, because the first stone of your work is home improvement through the use of talk and action. A consistent adherence to this point of view will bring to the forefront the role of family leadership in the forming of work-based aims. If these aims are formed not in thinking apart from acting, if they evolve from the process of acting, then the role of those participating in that acting cannot ignore your lead.

In a persistent act to practice doing things to learn from the experience, every effort ought to be made to set up a safe place in the school for learning the power of critically observing the growth of your child's own experience in terms of the goals in view, and to listen and learn who the key helpers are and how to lead them to experience these aims. Murphy and Alexander (2004) wrote about the role to persuade, as being the cornerstone of the education process. "There have been changes in that traditional condition where those who guided the education process—curriculum experts, philosophers—assumed the thinking for the learners and reserved the understanding of educational aims as their own exclusive privilege" (Taba 1932, 216).

Often the parent and child together, are not aware of the ultimate direction of the school in its lead role to prepare children for the workplace

(Cuban 2004). Without a focus on ways to set parents and their children up to experience learning by doing projects, it is a wasteful school system (p. 45). When you as the parent, decide to take action to practice moving your child through the school system, you replace the rote learning they call an educative process, with goal directed learning. When parents organize student study teams, they are able to observe teachers and their reports, listen for answers to where they are being led, learn from their experience in terms of the goal in view, help formulate a closer correlation with the actual child, and lead in the demand that the child be made more aware of the behavior in which the process affects.

Most often than not, the child and parent are considered to be naïve in the area of knowing what things they need to experience to make the educative process work for them. "This characterization of schools and leadership both describes reality and suggests a way of looking at school, especially secondary schools, which are places where the students are at once children and adults, dependent and independent, bursting with autonomy and hungry for adult affirmation and approval, confident-seeming, their talk more assured than honestly questing, but clearly scared about their futures" (Sizer 2004, 33). This makes the case, a truly participatory practice is needed not only to expand awareness of the child's needs in this day, but also the child's right to learn how to think through their participation in the act to learn in school.

Hence, the educative experience in terms of feeling the effects of school-based learning requires that the child be free of thought. In other words, the parent and child are supposed to be emotional about the experience of school in the deconstruction of such personal aims. These values for learning how to use the experience of education to create real results in the now, and not, in the near future, are resisted by the schools internal protocols. And because of this, the school is not set up to receive the child nor the parent, unless they are accepting the educational practices being offered to develop the process of that experience. But the process is neither concrete nor complete in the interests of developing the whole child (Cuban 2004).

What then is the role of the cognitive experience of school? What type of memory, and is rote memory, free of thought? Yes! This is what may take place; when parents can be convinced that their aims are naïve. The means of persuading the parent and the child to move through the school process free of personal thought is by arousing their emotions or feelings (Murphy and Alexander 2004). When the process of school experience is being driven through the emotions the parent and the child are more apt, than they are not, to lose signs of focus. That is a loss of physical discipline and self-control. Once the parent and child are in this state the school may then appeal to their need for a sense of reason. By design, school practices are set up to help them move through these feelings with signs of care to apply and adhere to its directives.

That means the parent and child is not suppose to evolve as a learning team, that is, from the experience of being schooled by educators, who do not care for family events, goals, and practices. And, it does not matter how closely these aims are related to the schools interests. The academic, social, and vocational curricula are limited to the schools sense of what it means to learn by doing work-based projects. This is a behavior that eschews cognitive constructs or mental processes. The reason, in which parents and children both are hurt and injured, is through such school-based learning and practices.

Chapter 4

A Brief Talk about Our Work

Education and Family Improvement

Being able to read and interpret your feelings as you move through puberty is important, but not to be able to sense, feel and focus the act during the ages of 19 to 25 is a break down. During this stage, young adults are required to know how to receive people as objects to be processed, as in learning to move through them with a focused state of mind. All people are required to sense and receive data, but during this stage an advanced level of cognitive and mental processing is required to interact with higher levels of formative states.

This transition from an informative to a more knowledgeable state give rise to the skills that underlie ones sense of self as a technology and/or human system science. "This transition and that from skill to know-how underlie the development of technology and of human technical culture" (Hall 1959, cited by Gaines 2004, 323). Young adults are required to know how to use the human system to produce new ways of knowing from their capacity and potential to move through emotional states using the sense and receive process cycles as tools and techniques.

Jimmy is a twenty-one-year-old young adult, whom has graduated from college with an open mind to learning how to live. This means that over the past twenty-one years, Jimmy has learned how to set his mind up to receive contact in home, school, neighborhood, and workplace situations. To live in a home with only one biological parent, to learn in a school with multiple ethnic identities, to think in a neighborhood with low achievers, and to respond in the workplace with the right state of mind has always been a challenge to Jimmy.

Jimmy came to Save Our Youth, the Next Generations to learn how to help improve learning and support services to children and families, as a participant in its leadership and child development plan. Jimmy wants to become a political leader. Hence, Jimmy needed to learn how to help the helper set him up for contact with self, other people and the environment as a learning and support service to focus his act to work with us.

We had to set up a path to receive information, a path to process knowledge, and a path to respond to the experience of learning how to practice helping the helper help, Jimmy, become more reflective. Since Jimmy came to us with an open mind to learning how to help, he has fit right into the study of the human information processing age. How he, Jimmy, receives contact, processes interaction, responds to cooperation, and acts to participate with an inner sense of self is in the aim to learn with other people to use the environment to think.

The Information-Processing Age described throughout this book how children, youths, and young adults have to be led to learn how the brain comes to think with a real sense of self, other people, and the environment in mind. Too often, a child may grow, a youth may mature, or a young adult may develop with no sense of how to use the brain to live in a home, learn in a school, to think in a neighborhood, or to respond in a workplace. Hence, he or she may lack a sense of how to lead a family, use an education, participate in government, or to focus business affairs.

Jimmy graduated from college with a 3.8 grade point average. Like many other young adults, Jimmy had done so without learning how to lead a family, use an education, participate in government, or focus his business affairs to live through the information, learn through the knowledge, think through the experience, and to respond through the reflections of being able to process contact, interaction, cooperation, and participation. In other words, Jimmy was not taught. He had a human system to help him learn how to practice, living each day to become more informed.

How can a child, youth, or young adult move through our public school systems without a sense of why the body is physical, the brain is mental, and that they both are products of the human system? Instead of graduating from college to go after a high-paying job with a leading human technology firm, Jimmy sought a basic entry position with the government. Jimmy had not sought to learn that at his age, and with his ability to read, write, use computers, cell phones and follow instructions, where his talents were in demand.

Clearly, Jimmy was not thinking as he moved through the grade school and college systems, how to use his education, to set up his physical and mental sense of self as products of his human system. In other words, every child, youth, or young adult will go through a period(s) of doubt or skepticism about feelings, identity and information processing. We found that Jimmy had this need to learn what am I, who am I, and how am I going to produce.

In this sense, it was important to lead Jimmy to learn to study what he thinks, who he thinks he is, and how he thinks it feels to chase the goal to become a political leader. The body, the brain, the human system: Jimmy's body, brain, and human system had to be set up for human systems research. Jimmy had to be led to learn about self, other people, and the environment from a Progressive Investing perspective; the act to live each day to learn how to become more informed.

Jimmy had to be prepared to work with children, youths, and young adults using a child-centered learning system. Hence, he had to be led to study the home, school, neighborhood, and workplace networks as places to plan, organize, and tests his feelings about becoming a political leader. In this same sense, he had to be led to see himself in the study of family, education, government, and business affairs as an adult learning system. In other words, Jimmy had to be led to reflect on his own feelings, while learning to respond to the needs of another person.

As youth and young adults move into adult roles, they need to be responding to their education as a practice. A practice in the use of the English language: to practice the use of reflection, to practice the use of identity, a practice of using thought, to practice the use of interaction, to practice the use of networks, to practice the use of feelings, to practice the use of sensation, and to practice the use of green technology—the human system. Jimmy came to us for help. He left with these skill sets.

Over a two year period, Jimmy was advanced twice. Jimmy is now a government program analyst with high aspirations for roles of leadership. Once Jimmy began to settle down and learn how to aim his sense, feel and focus cycle he became more aware of his brains capacity and potential to learn by doing things to practice moving through the experience. Many other examples are being recorded through our yearly events, where young leaders like Jimmy learn how to learn with us.

Jimmy's mother is black and his father is white. He never got to know his father. Hence, he was upset by a major life event. Not having access to his father, because he moved to another state. He grew up poor living with a careless step father in the inner city. His mother was very caring and nurturing. He was the oldest child, by nine years. He grew up with a younger brother and sister who did not get along with him, because he was treated differently by his step father. He had issues that we had to work through.

From age nineteen to age twenty-five, young adults that have been hurt by major life events are more susceptible to feelings of anger and pain that cause him or her to resists moving into responsible roles of adulthood. The SOY Family Leadership Academy was established in 2000 to help move children, youths and young adults in crisis through the learning system more prepared. We call this process learning how to live (Slaton and Slaton 2002).

This process will work for you, because of the things we do to prepare your brain and body to learn how to live with self and other people, to learn how to learn from us. For instance, you need to learn how to process more information to meet your need to become informed. While we will focus on family, we work through your basic needs as a person. First you will be set up with a mentor. A mentor is a person that works with you to enhance the chances for your successful completion of the Leadership Academy (p. 21).

The mentor and you are assessed in the following core areas:

- Self-Development
- Information Processing
- Participation
- Community-Based Learning

The first step is to set up your sense of self through your understanding of family affairs. You will work to gain the status of a "family advisor." This is the role of a person who advocates for things that strengthens the quality of life. You will gain advanced knowledge of family systems and ways to prepare self and other people for contact with parents, children, and designated close relatives that work together to reduce the threat of family decline.

Specific areas of learning may include the following:

1. Family Leadership

    - Male and Female Relationships
    - Child Development Costs
    - Family Business Plans
    - Personal Credit Reports
    - Family Income/wages/earnings
    - Family savings accounts
    - Family checking accounts
    - Family Monthly Budgets
    - Family Resources
    - Family Home Ownership/Rentals
    - Family Allowances
    - Family Mission, Goals, Objectives

What is information processing? Information processing is what your brain, body, and senses are designed to do to move you through contact and interaction with self, other people, and the environment to learn how to live. You will learn how to practice living each day to become more informed, as an information processor. An information processor takes in sense data or information from other people through the sense, feel, and focus cycle to receive, process, and respond mentally to gain knowledge from the experience.

Specific areas of learning experiences may include the following:

2. Progressive Investor

    - Planner
    - Organizer
    - Coach
    - Learner
    - Worker
    - Leader
    - Manager
    - Executive
    - Helper
    - Developer
    - Entrepreneur

What types of participation patterns do you want to build? Participation is what you will need to be an effective family leader. People have to feel their need to follow you, your advice, or your direction. You will learn how to think things through as a participant. To participate as a participant, you will need to learn how to take action and to learn from the experience of contact, interaction and cooperation.

Specific areas of learning may include the following:

3. Participant Observer

- Motivation
- Purpose
- Values
- Communication
- Discipline
- Self-Control
- Judgment
- Choice
- Flexibility
- Practice

What type of community do you want to learn how to live in? Community-based learning prepares you to learn how to live with self, other people, and their environments wherever they may be in the world. You will be prepared as a community-based learner to practice developing a higher sense of land-based systems.

Specific areas of learning may include the following:

4. Community-Based Learner

- Progressive Investing
- Human Systems Research
- Home-based Learning
- School-based Learning
- Neighborhood-based Learning
- Workplace-based Learning
- Living Systems
- Learning Systems
- Thinking Systems

- Response Systems
- Feeling Systems

From the business perspective, you will learn that "space is never empty. If it is filled with harmonious voices, a song arises that is strong and potent. If it is filled with conflict the dissonance drives us away and we don't want to be there. When we pretend that it doesn't matter whether there is harmony, when we believe we don't have to walk our talk, we lose far more than personal integrity" (Wheatley 1999, 57). We have lost our point of focus. You move from a sense of self, to a sense of self and other people to learn how to be an inclusive thinker.

Your future and the future of your children depend on the steps you take to prepare safe spaces to learn how to set up a family as a learning system. As young people, you have a duty to self, other people and the environment to act quickly and wisely when you begin to feel or think you have a need to date or experience sexual interaction. To do this, depends however, on the state and condition of your feelings for self and your thoughts for other people through your sense of family as a value.

What kind of family do you want? Do you want a working family or a broken family? A broken family struggles with issues of identity and purpose. What type of children do you want to rear? Every child you help bring into the world deserves a chance to learn how to live with you. From your resources, what type of human assets do you want to grow from infancy through adulthood? From your earned income, what type of roles do you have that offer the experience of wealth as he or she learns how to live with you?

Is there a need for wealth in your family? In response to the challenge of creating safe home, school, neighborhood, and workplace networks for you to experience, does this need apply to your sense of family? Is being able to protect and serve your family important to you as a mission, goal or objective? Where do you learn these things? Today for instance, right now, can you write a family mission statement? How fair a person, are you? Are you a person of thought? We want to know that you can address these basic challenges related to how you may become a responsible adult and family or future family leader to help you in your effort to focus the way you respond.

By the way, a family leader is an executive—through perhaps misfortune—children depend on their parent's capacity and potential to serve in the role to provide him or her with learning and support services.

What have you done today that might qualify as a vested interest in the value of becoming a leader and executive over the affairs of self, other people, a child? Learning in this sense, is the aim to gain the experience needed to succeed. Hence, learning how to live is a lifelong aim. To explain this special connection between how we live and how we learn from experience it was necessary to outline the terms noted above as objectives.

Living through and learning from experience . . . the act to lead in the act to execute . . . a family business plan requires a system. A system is a process that allows data and information to flow through specific events that can be aimed to reach certain goals that produce meaningful outcomes. This means, you should not start a family, or, you cannot effectively develop a family with less than an "on-hand cash balance of $5000" or more. This is start-up capital set aside to back your investment.

1. A more precise statement: Because you need to cover family, education, government, and business expenses, as well as transportation, housing, food, and other basic needs, you need a minimum on-hand cash balance to start a family and a reliable source to acquire new funds. Hence, your goal is to live and to learn through focus.
2. A clearer statement: People who grow up in America and want to learn how to compete must experience discipline and control over their spending patterns to lead and execute a family business plan. The way you prepare is the way you practice.
3. A feedback statement: People notice you have a capacity and potential to focus how you choose to live and learn from experience in connection to family, education, government, and business systems. You feel things to think through the things you feel.
4. An independent statement: Something inside of you (character) moves you to take action to be seen as responsible to self, other people, and the environment as signs of knowledge, judgment, and expectations. You lead to model a response pattern.
5. A consistency statement: Knowledge, judgment, and expectations are experienced with the buildup of confidence along a continuum where you have learned how to process tasks and business values. You have focused your aim to live a good life.

Learning from experience is personal to organizing a sense of family matters. You have personal goals, and then you have family goals. You set up the personal goals to relate to your family goals as a business venture. They then build connections to anchor how you feel and think through your needs. Learning how to feel for your needs and learning how to think through the experience of your needs is the way you protect and set up a safety net. If you go back and review the objectives we have set up above, study the words. Hence, they become part of the experience.

You are learning how to look at the words to experience how they feel, but in connection to ways of knowing how to use them. Keen and Staples (2003) describe the meaning of knowledge as a complex act in the process of experience. "The basic economic resource—the means of production—is no longer capital, nor natural labor. It is and will be knowledge" (p. 21). You live through the process of thought. Thought is the experience, and knowledge is the result of processed thought.

1. You feel the words through eye contact. Talk through the words advances thought.
2. Think through words. Talk through the words to a line of focus.
3. Act through how the words feel. Talk through the feelings of the words.
4. Reflect on how it feels to think through words. Talk through the words you know.
5. Define how you act, feel, and think through the words. Talk through the words as an outcome.
6. Write the words down. Dump how you use to feel these words.

Each time you feel for a sense of the word, it is new. You learn more and more each time you focus. You use words to learn how to act, feel, and think through your feelings as they grow to be more known to you. You live, learn, think, and respond to growing more aware through your senses, body, and brain. You experience ways of learning as ways to move through paths of focus.

The Scientific Context of Parenting Children in Crisis

To many of us, parenting is a role. But what is in the role of parenting? Do you have to care? What does a parent have to care about? To us, a parent has to feel the role to care about the life of a child. What signs of care does a parent have to show? To us, a parent has to show feelings, and what they mean depends on how it feels to focus a sense of control in the role, of which it is applied. To parent the life of a child: Each stage to grow, to mature, and to develop has its own unique characteristics that need to be set up. To grow with a child, it is a parent's role to learn.

In the case of having feelings, a parent must learn how to move through them from abstract positions to more practical positions. In the specific case of the applied use of caring and feeling for the child as a parent, it means to be a human developer. This set up has not always been observed, because of the criticism a parent may receive when they act through feelings of care that are not yet clear to them. The intimate connection between feeling and parenting has to be grounded as signs of care about the role of, and the parent has to relate to the act as a tool, to learn the demands of the craft beyond theory.

But as a core stone of parenting, feelings have to be qualified and tested through the events of becoming real and less abstract about the positions taken to inform self. Wherein, the risk of being criticized in the effort to build up an approach to understand the role has to be learned from. The results of such an approach will be held to reason, as the feelings being exposed are moved from within oneself to the world outside of oneself, through the use of talk. In the act to set up the role, is the act to act in the role itself. Parenting is a science, and talk is a tool. The first step is to learn how to use talk as a method.

How we talk to a child is very important to how we talk to self. Is this not true? Now is the time, look at the way you as a parent receive, structure, and express how you feel through the use of talk. The search for the true path has been one of the main aims of parents wanting to learn through parenting as an event, how to help a child that is hurt. In this sense, the capacity and potential to feel your way through the role of parenting has to be set up for such a call to action. Experience, in learning from the event moves the feelings of parenting to signs of focus, where discipline and self-control take hold of the act. How we use talk to focus brain and body events is an intelligent response. Hence, the act itself is sent to memory.

The result has been that once the act is experienced in the role to develop the child's connection to states of becoming, parents learn from their sense of feel for the use of talk. Perhaps in times past, this method has been blurred. In our view the parent is now set up to feel and care for the experience in giving of oneself to a child, for the sole purpose of valuing intelligent reflection. Taken as the causal aim, to learn how to value events is to become more mindful, and then in acts to feel and care there will be signs of an intelligent parent, reflected. Talk is regarded as important to the path of focus, which is to build a capacity and potential for helpful refection.

Hence, the science of parenting is a physical practice. This cognitive experience is set up by the particular method used to send feelings forward and receive feelings backward. The transfer of the method from within oneself to the world beyond oneself moves from the abstract to the concrete sense of self. How do we treat a sense of self, when the feelings form and mould signs of care in the role of parenting? The effects of such an event causes the parent to feel more real, because we care about being physical or how we are treated with signs of care. We can make sense of the use of talk as a core practice.

We can observe through the use of talk and action how parenting unfolds as a humanistic and social practice of becoming reflective in the study of the child's response. This step is undertaken to plan and execute what parenting means and why the event is physically connected to doing things. But the method of doing things is connected to the child, in the sense of, acting to learn from the experience to the effect of the experience. Hence, we can feel the parents act as the child learns from the experience as feelings and the observation is thinking through. We think signs of care are what lead feelings to enter a reflective state due to the practice of the role to observe the child.

It is true then, that the state of parenting in the humanistic and social sense is physical by nature and intellectual by cause and effect. We will experience learning in the act to live with signs of care due to the complexity of feeling things and thinking through the things we feel, the child's need to be led as and treated as an emerging technology. Parenting then, is a humanistic effort to construct ways to learn how to care and feel more capable in social places. Parenting, in its narrower sense, consists of certain aims in the use of words and talk to make contact, interact, cooperate, and participate in acts to live, learn, think, and respond to the people in their child's home, school, neighborhood, and workplace network.

Behind each event is a parent's physical practice, to aim the act of thinking through the humanistic lens of parenting, to talk a certain way or use words that relate to a certain philosophy from the learning method.

Control and discipline, brain and body organize the capacity and potential to learn through the method itself. Hence the science of parenting entails the control and focus of feelings about how it feels to receive talk and words and its relationship to care. Care has certain effects on the act of thinking through the humanistic lens of parenting as it causes us to reflect, relate or represent a learning method.

Many of these feelings, if they are seen as caring, would be recognized as signs of becoming more mindful through the learning method. The basic theory here is that by being aimed and focused will help the transfer of feelings to be recognized. The learning method is open to signs of care that are seldom brought to light in the aim to feel the transfer of feelings. And, as feelings become more focused they influence how we live through observable objectives. For instance, the body is a physical act to observe, and the brain is a practice objective to sense, feel, and focus feelings and signs of how it feels to care.

As observable objectives, resistance is to be expected, because as we become more informed in the role of being a parent we also act more mental than physical. In this case, people can see us feeling things out in search of signs of care to focus the transfer of feelings through the learning method. In other words, parents compete for prestige through the use of methods such as these, so their child and other people will see them as being important. Hence, in social places, we are observable objectives to accept or resist, since we feel the role to care and build intelligent reflection. If we believe in the method, it is because we are learning from the act to experience the event, which informs us.

Is this not the bridge of conscious thought? Do we want a practice that informs our consciousness? So much of parenting is blocked by resistance to schools of thought that promote the role as being an art and science. But it is what it is. The danger of becoming a more informed parent may be seen as a threat to teachers and other child care workers, since they benefit from the parent being less skilled in the areas of parenting and teaching. Hence, the role of a parent is to lead. To us it is beyond doubt that many of the roles of parenting relate to entrepreneurship, where certain skills need to be structured.

If you choose to follow this path of observable objectives, it leads to informing consciousness. Who are you? What are you doing? Why are you doing the things you do? Where does being you make sense? When is the best time to be you? How can you help? We live each day to learn how to become more informed. Hence, the science of parenting a child in crisis starts from where you are . . . in the role to feel and care about the event . . . to act as a parent. Parents have a philosophical point of view that has to be built up over time in the act to advocate for the whole child to be allowed to grow, mature, and develop as a person.

To the Student, Learner, or Participant of Progressive Investing Practices

Special Theory
Progressive Investing is a special theory of purpose that we apply to the context of Human Systems Research. Hence, the linkage between word use and action sets up several process functions. Such as, sense perception, the receipt of sense data, and the process pattern of which each function is a special element in the act and practice to become more informed. Theoretical work of this form, emphasize the value of being open to the study of words and action to learn from the experience of an aimed state of mind.

Special Practice
This emphasis then is on the practice that fuses both the physical and mental acts of participating, wherein which, contact, interaction, and cooperative states are sent forward or received backward and the special practices linked in doing this work are framed. In our work with parents and children in crisis, special practices like Progressive Investing is needed to study, to understand, and to explain events. Hence, the purpose of pilot studies using a laboratory as a qualitative method to build up experimental Human Systems Research is to feel for a sense of cause and effect in past, present, and future events.

Special Acts
Living each day to become more informed forms the special acts to live, learn, think, and respond to tension, stress, and pressure. The words *Progressive Investing* mean to live with this purpose in mind—to act through living, to act through learning, to act through thinking, and to act through responding to circumstances. Educating the brain of a parent and child that have been hurt by major life events require special acts to lead them to realize they have a brain and how the brain is used to help them live, learn, think, and respond.

Special Science

It is the point of this book that the brain, body, and senses form as the core elements of a person's human system. Hence, learning how to live, learn, think, and respond to contact is a human systems function. It is useful then, to study self, other people, and the environment to learn about the things the human system can and cannot do. The study of self, other people and their environments is called Human Systems Research. The human system and Human Systems Research are distinct both as a body of knowledge and as a certain way to gain experience through the process of knowledge.

The Slaton's introduced these terms (Slaton 2009), to explain how the body lives, how the brain learns, how the human system thinks, and how human systems research responds. Some examples are as follows:

1. In what way do we use the brain to learn?
   a. We use the brain to receive things as a system or processor.

2. In what way does the body live?
   a. The body lives to feel things for feelings.

3. In what way do the senses work?
   a. We use the senses to send data to the brain.

4. In what way does the brain and body think?
   a. The brain and body acts like a system to think through controlled feelings.

5. In what way do we study brain, body, and sense events?
   a. In Human Systems Research the brain, body, and senses lead to the study of self, other people, and their environments.

The parent and child's human system is defined by the way he or she interacts to live, learn, think, and respond as a set of elements. As these elements are being made more aware to them, it is the way we structure these words in relation to home, school, neighborhood, and workplace learning that makes up the special science. One is a human learning science, and the other is an environmental learning science. These may be referred to as human systems science, to distinguish them as being interrelated as man-made events and authentic sciences.

Special Places
Human systems researcher's learn how to study action as a goal, which then builds on the relationship between the practice of Progressive Investing and the skills needed to improve a parent and child in crisis at home, in school, in the neighborhood, and in the workplace as special places. Parenting and growing up, for a parent and child that have been hurt by problems in these special places make up a field of experiences to sense, feel, and focus as they make contact with other people who share them. This makes them special places to live, learn, think, and respond to compete for status, wealth, and resources. Hence, they are special places, as symbols of how we live, learn, think, and respond as a society.

Special Education
Many parents and children in crisis have to learn how to move through day-to-day events that are threats to his or her physical or mental well-being. The problem is that so many may feel ashamed of being upset that they resist the need to deal with the events that place them at risk. This is a special education problem, because of the way being hurt may affect language and cognitive experience, in addition to physical identity and the mental process. If therapeutic, social, and academic processes are used to assess how the parent and child feel through events, then more intelligent family-centered learning plans can be set up to help them move through the crisis.

Safe Places

A child's "human potential is vastly underestimated and an unexplained territory" (Maslow 1998, xx). The home has to be a safe place. The home is the first place children use to practice learning how to feel things and connect in the process to think through, as they are able. The healthy child can be expected to be flexible and realistic; that is, able to shift from growth to defense as circumstances may demand (p. xxiii). The contrast in the motivational life of a child that is growing up hurt is between how they are motivated to grow and how they are motivated to defend their emerging sense of self.

Maslow's motivational hierarchy of needs is physical, mental, social, environmental, and emotional needs to compete for space, time, and energy. For instance, a parent and child needs to be able to move through home, school, neighborhood, and workplace networks during periods of growth, maturation, and development in acts to sense, feel, and focus how they live, learn, think, and respond to their needs. If not structured appropriately, unmet needs can create fear and anxiety problems that affect a parent and child's sense of safety, when they are not set up to compete for space, time, and energy.

This is why we build systems, strategies, and process cycles for the parent and child to experience along a continuum

Progressive Investing Model 9
Assessment (Slaton, 2009, p. 34).

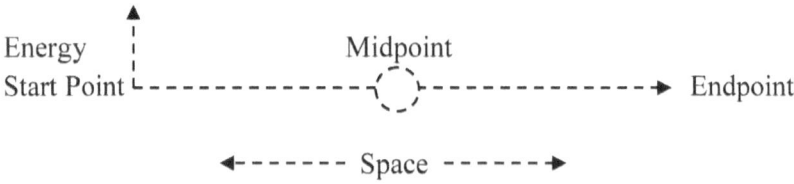

Learn how to live each day in the act to sense, feel, and focus how you live, learn, think, and respond to the way you move through home, school, neighborhood, and workplace networks with the intent to grow, mature, and develop. Feeling safe is the cornerstone of the effort to live through each act. Hence, a great way to start addressing the challenges posed by confusing systems, strategies, and process cycles is to structure the way they will be experienced over a certain time frame.

Building up a parent and child's sense of self, other people, and the world around them should help improve how they feel their need to compete for safe places to live, learn, think, and respond at acceptable levels. It is recommended that the system we have set up be used as a way to practice learning how to live in a home, school, neighborhood, and workplace network to compete for status and attention. This strategy will help to ensure that a line of focus is set up to move the parent and child into the school system, feeling safe and more open to contact.

It is vital to a child's entrance that parents actually take their child to school and introduce themselves to the teacher(s), with their child in hand. Eye contact is critical in the assessment of parent, child and teacher feelings. There are not substitutes for setting up process cycles between the parent, child and teacher to study signs of care, and to ensure the child feels connected to a good behavior from the start. In a safe and structured process cycle, the teacher is more capable of informing and instructing the child on ways to learn how to learn from them and with them.

Setting up neighborhood experiences are more difficult, because there is less of a structured path for people to make contact and interact with care. Many parents and children that are hurt experience discipline and self-control issues in their neighborhoods than in any of the other places, because of what we call free space. When there are less structured events in a neighborhood for the parent and child to experience, the lack of controlled space can create a great number of problems with discipline and self-control.

The neighborhood is the least supervised space. If the parent and child do not feel their need to act through signs of care for who and what they represent in the physical and mental sense, then they are more likely to behave in a patterns that reflects this disconnect. In addition, parents and children who lose their sense of focus during the developmental years of a child's eligibility for a free and appropriate public school education can be disconnected from their family, friends, and peers as being too toxic to live around. We relate to this as being signs of family decline, school failure, delinquency, poverty or lack of employability.

Many parents and children in crisis live to ignore the helper's advice, and then later, they come to realize the value of it. The home, school and neighborhood are their workplaces. They need to live each day to learn how to feel them, through greater signs of care. One way to structure neighborhood life is get to know the people that require parents to parent their children and children to act like children. This means that the parent and child are working to create and to manage physical and mental signs of care through their feelings to discipline and control their thoughts.

In the end, feeling safe really depends on the parent and child's motivations. Are they here to help, or are they here to spread their pain? The home is where signs of care create safe feelings and improves as the parent builds up an effective strategy to lead and develop the child's sense of self, other people and the world. This way, the child will leave home each day, in state of mind to learn. Hence, the parent and child will use the home as a safe place to live.

Parents, children, teachers, and families can work together to make our homes, schools, neighborhoods, and workplaces feel more safe. Here is how:

1. Use our space to learn how to feel your space to help set up a defense against family decline, school failure, delinquency, poverty or lack of employability.
2. Use the safe place process to define the human services you need to improve the delivery of learning and support.
3. Use the ideas that come to light to study the social worker, nurse or doctor, educator, employer, child developer, and other professionals to learn how to connect your system and strategy to the people that can help you structure the body of knowledge required to improve your quality of life.
4. Use today to present enough signs of discipline and self-control to focus your work on the safe place process.
5. Use this system of knowledge to recognize the human systems that connect the people to your work.
6. Use your brain, body, and senses to achieve this.

Parents and children learn how to use their brain, body, and senses as a human system. The human systems model draws attention to the dynamic patterns of relativity between the person, the people, and the whole human race and the context, the history, the earth, and the global world. In its essence, the human systems approach is a study of how we feel things, objects, and environments to think and reflect on the experience of moving through these feelings.

Parents and children, working within this aim, seek to focus brain, body, and sense events through, and in the safe place process emerge feelings for self, other people, family, things and objects. Thinking systems or the practice of thinking through the things we feel: The systems thinking model, validated by Checkland (1998), is applied here as a human systems feeling approach (Slaton 2002) to structure thought through practice.

If the parent or child does not feel safe in their space, then the experience of self and the experience of other people will reflect levels of fear and anxiety. The first experience is raw feelings; the second experience is how they feel the things they can. In that experience of experience emerges feelings of thought from the act to receive the experience. Each step has to connect to the next, which creates the relativity between two points. The affects of which can be identified by an upside down pyramid extending from the midpoint downward.

Progressive Investing Model 10

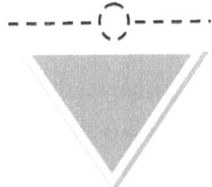

According to Einstein (1961), the pull between the two points are progressive or regressive. Hence, the safe place process starts from the way the parent and child is set up to practice the sense, feel, and focus cycle to move across the scale. We have learned from our experience and the experiences of participants that input is resisted when the senses are used to think rather than feel for, which narrows the path to receive the experience. "Systems thinking then refers to the use of the brain to recognize, think of, and form the coherence of the wholeness" (Anderson and Carter 1990, 9) in the experience of the process of total thought through the peak of reflection.

On the base, the experience has connected feelings to elements of thought. As the parent and child's act cause them to move toward the midpoint, the elements of thought represent stages of intelligent reflection. It is the act to grow that causes the parent and child's effort to be more capable of being understood. Hence, the need for safe places is accepted as a viable defense against feelings of fear and anxiety brought on by internal conflicts between self, other people, family, events, and environments at odds with the flow of sense data to the brain in the process of thought. This means, a sense of feel for feeling safe in home, school, neighborhood, and workplace networks has to be focused through the practice of thinking.

You are in Crisis if you Use Street Drugs

The word *street drug* refers to an illicit substance that is unlawful to possess or to use, because it has been manufactured for sell or to be illegally distributed using neighborhoods, cities, and streets. The parent or child may not really want to use street drugs, but because they live in a certain neighborhood, they are targeted and may do so to please others. Once the parent or child start to use street drugs, the capacity and potential to stay focused is disrupted. This is because the mental process becomes compromised as the need for using the drug, obtaining the drug, talking about the drug, or trying to find paraphernalia to use the drug takes them over.

Street drugs are responsible for substance-induced disorders that affect behaviors in the home, in school, and in the neighborhood (Slaton 2006). Thomas Sazasz (1990) wrote about substance abuse disorders as an illness long ago. The buyer is sick because his or her behavior is not intentional, but the behavior of the seller, street drug dealer, is a deliberate and involuntary act to push a person into a pattern of drug abuse (p. 218). If a parent or child in this situation fails to function and perform in structured home, school, neighborhood, and workplace settings, drug use is a threat to his or her well-being.

This is evidence of why you should fear drugs, and it is everywhere. People at the bottom of the social structure become the underclass. People above the underclass become worse off as the threat of drug abuse overwhelms their thought patterns. Gangs become more known to the street drug user. As a neighborhood is taken over, random shootings and acts of violence occur. People who either sell drugs, use drugs, or offer drugs for favors practice tricking people into trying drugs. A drug dealer or person trying to trick you may substitute one drug for another more potent one to increase a certain feeling and mood change that affects the senses.

The physical damage to the body may include the way a parent or child becomes a sign of distress. The body develops a physical need for a certain street drug. Hence, the image of being enslaved to some form of substance becomes more apparent as the user looses the capacity to control their senses. Physical substances injected or ingested into the body damages brain connections through sudden changes in moods that affect mental processes. For instance, the way you sense, feel, and focus to receive is altered by the deviant behavior or abuse of the body and brain. This may be thought of as an attack on the central nervous system, where mind-altering effects like this lead to physical signs of damage.

The cognitive experience of street drugs affects emotions, which in turn affect attitudes (Ray and Ksir 1993; Kendall 2001). Street drugs stimulate feelings of excitement, relaxation, power, and control through mood swings that trigger levels of need for the experience. Since drug abuse is physical, mental, and social, it causes problems as the central nervous system becomes more susceptible to signs of withdrawal being linked to the frequency of drug use that forms the addictive pattern. The user does not feel the same kinds of emotion or attitudes when the drug is not being used. Hence, the nature of the chemicals is designed to trigger depressive states.

Say no to street drugs. This may lead to a distorted sense of identity. You know you are not the same person. You care, but you do not have the same values. You care, but you cannot sense and receive the need to quit. In other words, you care, but you do not want to stop your use of drugs. You care, but you do not feel for, see, or hear the people you hurt. If you can say no to street drugs, then we can help. If you use street drugs, you have to stop. If you like the way street drugs make you feel, you are in crisis. Street drugs are designed to change how you feel things. You change the way you show signs of care for life each time, each day, each week, each month, and each year of street drug use. Street drugs are used to trick the brain.

Street drugs kill dreams. Street drugs control your lack of treatment intent, ego needs, and foster dependency on drug dealers. Say no to street drugs for the reason that they kill your family affairs. Street drugs destroy how a parent and child live. A parent and child are less able to function and work in their home, school, and neighborhood networks. Moreover, unborn children are born addicted through parents to street drug use. Hence, a child's drug addiction recycles the medical problem. You have to want to change the things you can. You have to want to accept the need of help to stay away from street drugs. Street drugs kill brain cells.

The way we help is by setting up ways for you to express yourself while we talk to your brain. The system, strategy, and process we set up for a parent and child in crisis to experience coincides with the learning and support services we use to improve their home, school, neighborhood, and workplace networks. Evidence of drug abuse is very clear to us. There are signs of harm to self, family, other people, and their environments as forms of physical, mental, and social trauma. This means that the entire circumstances that surround the effects of drug abuse are more likely to be full of drama—rants, pain, and hurt.

Hence, the physical, mental, and social basis for treatment has to be set up to reduce family decline, school failure, delinquency, poverty and lack of employability. The focus is on the child. We talk to the parent's brain. The parent learns to journal, and we lead the parent to talk to the child's brain. Compulsory Building Human Asset Meetings are used to learn about issues of tolerance, withdrawal, and intent. Family-centered learning is set up to offset problems like having to isolate and withdraw by providing staged social events. Talking to the parent and child's brain allows us to structure the approach we use to learn how they each function as a human system and living organism.

We refer back to Szasz (1990) on what is a proven brain disease: "The brain is an organ and diseases of the brain manifest themselves, inter alia, as disturbances of behavior" (p. 49). If the parent or child cannot sit still, aim their head, control their body, and use talk with care and kindness, these are signs of such an effect. Although Szasz goes on to point out "No one is compelled to use street drugs and that therefore, in the final analysis, people take drugs because they want to take them; and although it is also obvious that there are many complex cultural, religious, and legal reasons for the so-called drug problem; it was fashionable to attribute the problem to the drug abusing person's mental illness" (p. 292).

We treat the behavior by talking to the brain to learn how open the parent and child are to acts to live, learn, think, and respond to change. People who act crazy may not be crazy, but are acting crazy. A behavior is a fixed state of responding. Hence, when you act to change a fixed state, it is to alter a cognitive state or experience. People who act like they want to be sane may not be sane, but learn to behave like they are sane. The treatment goal, then, is to act with signs of care for the result, which is to improve the ability to function physically, mentally, and socially. Literally, an attitude is changed through acts of character—to feel the physical, mental, and social affect of the work to focus—the act to live, learn, think, and respond in home, school, neighborhood, and workplace situations.

In appropriate use of the brain produces damage to the human systems ability to function, due in part to the behavior, tension, stress, and pressure on the body to perform in a certain and more noticeable way. When the human system begins to fail, it is because it has been locked into a fixed state of responding as a self-destructive mode of performing. In other words, according to Szasz, the addictive behavior causes the parent and child to grow into stages of insanity as the brain damage progresses as signs of mental illness. Being mentally ill refers to the physiological and psychological need to care for the behavior to avoid the painful symptoms of withdrawal (Ray and Ksir 1990).

As the Progressive Investing Model below indicates, the behavior is the event that has to be dealt with through treatment as the goal. Even if the parent or child in crisis does not reach the goal the affect between the aim and the focus produces an effect, which acts as a response to the event and goal.

Progressive Investing Model 11

Event	Affect	Goal
Behavior	Response	Treatment

According to Einstein, the pull between these two points produce signs of relativity. Hence, the affect can be calculated as an effect based on the compilation of sense data across the continuum of action. According to Maslow, the need for a sense of identity causes the pursuit of the goal to release a competitive response.

Progressive Investing Model 12

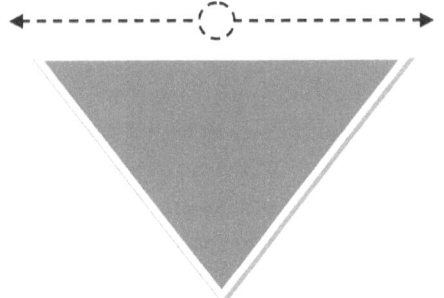

1. Feelings displayed in relation to home
2. Feelings displayed in relation to school
3. Feelings displayed in relation to neighborhood
4. Feelings displayed in relation the workplace

Each response is assessed in relation to acts to live in the home, acts to learn in the school, acts to think in the neighborhood, and acts to respond in the workplace. Changes in behavior are related to changes in the home, school, neighborhood, or workplace network where the parent and child are learning how to accept help from other helpers.

The social role of the parent and child as participant observers move the event to action, which plays a pivotal function in the aim to sense, feel, and focus language, cognitive experience, physical identity, and the mental process on the goal state—to change the behavior through new signs of personality, character, and attitude. "Roles are readily classified according to the social context or situation to which they belong" (Szasz p. 27). For example, the roles of home and life inside the family, roles of school and learning through education, the roles of a neighborhood and thinking through government, the roles of a workplace and responding to business as these social functions imply.

In Human Systems Research, the fact is "social research is the way we study the self in relation to groups: basic contact, basic interaction, basic cooperation, and basic participation in the field of experience" (Slaton 2009, 146). Among the basic functions of social research is the taxonomy of roles for the participant. Functions of identity classification are as follows:

1. We identify the comprehensive pattern of behavior to be changed.
2. We identify the comprehensive and flexible personality that is expected to emerge.
3. We identify the comprehensive and flexible character this is expected to anchor the process.
4. We identify the comprehensive and flexible attitude that is needed to represent physical and mental signs of control.

Hence, the fact that each role is socially identified to be cognized in relation to the treatment, together with the fact that each role can be played by multiple individuals generates social knowledge from each basic exchange in each act—sets up a strategy to prevent, intervene, and enforce problems, issues, and concerns across the continuum. Terms such as self, other people, and the environment all apply to the roles of social research in the sense of how we work to improve a behavior, treatment, and response.

About the Facilitators

We use information science models that allow us to set up Human Systems Research through the study of human-to-human contact. Trained trainers act as facilitators of a Progressive Investing Curriculum designed to

1. motivate strategic action, contact, interaction, cooperation, and participation;
2. improve the delivery of learning and support services in the home, school, neighborhood, or workplace environments by offering therapeutic services;

3. improve leadership and development through Human Systems Science;
4. set up a holistic atmosphere for individuals, groups, and organizations to experience learning through the use of open spaces, conference rooms, meal services, quiet reading rooms, hot tubs, and swimming for as a more reflective place to practice; and
5. prepare people for more effective roles in the workforce through their successful interaction with Progressive Investing Literature to improve how they live, learn, think, and respond to tension, stress, and pressure.

Dr. Christopher K. Slaton, EdD, is a human learning consultant and has served Save Our Youth the Next Generations for over fifteen years. He has advanced preparation in education, leadership, and human systems science that allows him to consistently display an ability to learn, lead, and implement processes to meet the functional needs of children and families in crisis. He is the author of *Education and Science*, a resource book for how the body lives, the brain learns, the human system thinks, and human systems research responds.

Dr. Slaton has designed Parent Education and Resource Coordination Services (PERCS) to help children and families in crisis gain access to the right family, medical, educational, mental health, social work, and corrections professional to reduce the risk of cyclic declines due to learning, behavior, and conduct disorders that were not diagnosed in accordance with a child, youth, or young adults years of eligibility for a free and appropriate public school education. To change the factors that contribute to an inability to learn, behave, and control one's conduct, he or she has to become more informed about how to use a home, school, neighborhood, and workplace network to focus how they grow, mature, and develop in these places. This is why in his book *Education and Science* (2009), the process is described as a green technology. The participant learns how to improve their home, school, neighborhood, and workplace networks as a human asset. Hence, this is called the Building Human Asset Process.

Through the Progressive Investing Institute of Focused Learning, Dr. Slaton has developed process cycles to explain how children, youths, young adults, and parents in crisis live, learn, think, and respond to tension, stress, and pressure. Hence, Dr. Slaton talks about ways to sense, feel, and focus your brain and body as a human system.

Dr. Dolores J. Slaton, EdD, is the director of Family Services and Lead Administrator in the Save Our Youth, Family Leadership Program for Save Our Youth, the Next Generations. Dr. Slaton earned her doctorate at the Fielding University of Santa Barbara as the first student of Progressive Investing. Her areas of focus include family, education, government, and business systems.

Dr. Slaton also works full-time for the state of California. Serving in state government for over twenty years, Dr. Dolores Slaton consistently demonstrates her ability to discharge the responsibilities required of leaders. Having a graduate degree in education, leadership, and change enhances her ability to plan, organize, and implement processes to meet operational needs within family and business networks. Using standards and principles of project management, coaching, mentoring, assessing, managing, and mitigating risk, aligned with business practices, have led participants under her purview to consistently achieve organizational excellence individually and as a group. Effective communication and interaction with internal and external family and business leaders, legislative policy makers, and other significant stakeholders has led to resolution of complex problems within her scope of influence, particularly within family systems.

Learning and Training

In a February 2010 Building Human Asset Meeting with State Assembly Member Jim Beall Jr. N. Kathleen Finnigan, legislative director, and Monique Ramos, legislative aide for Assembly Member Tom Torlakson, we discussed ways to improve learning and support services to our public schools. We need partnerships with policy makers, the State Department of Education, and the Los Rios Community College District for a chance to improve the learning and support services of poor, minority, high-need, and special-need children, youths, and young adults. Through the use of Family Leadership Academies as laboratory schools, parents become informed through the act and process of learning how to live in a home with a child that is growing up hurt by major life events in the home, school, neighborhood, and workplace network.

> Children that are growing up hurt learn how to live in a home; youths that are growing up hurt learn how to learn in school; young adults that are growing up hurt learn how to use a neighborhood; and adults that are growing up hurt learn how to focus on the workplace. Parents, teachers, students, school principals and other helpers make up small learning communities creating safe places for children, youths, and young adults that are growing up hurt to practice interacting with self, other people, and the environment.
>
> This strategy produces more effective workgroups, helpers, and human relations. Over the next six months, we will be attempting to bring this plan to the citizens of California, but with your help; we can bring this plan before the Senate and Assembly, the Department of Education, and the Los Rios Community College District. (Slaton 2009)

"As human beings experience the unrolling flux of happenings and thoughts which make up day-to-day life, both professional and private, they are all the time likely to see parts of that flux as 'situations', and certain features of it as 'problems' or 'issues.' These concepts and this kind of language—of 'situations' 'issues', 'problems'—are very commonly used in everyday talk" (Checkland 1999, 28).

On March 18, 2010, we informed Board of Supervisor Don Nottolli we would like to hold a Building Human Asset Meeting with Ann Edwards-Buckley, director of the Department of Health and Human Services for the county of Sacramento to discuss our work with children and families in crisis. We looked back on our previous BHAMs with Board of Supervisor Don Nottolli and hoped he would realize the importance of our work with children and families in crisis. Hence, we sought to use to use these meetings as a point of reference.

On April 15, 2010, we met with Ann Edwards-Buckley, director, Department of Health and Human Services in the Building Human Asset Meeting. The purpose of our meeting was to discuss how PERCS can provide new and innovative options for improving the home, school, neighborhood, and workplace networks of children and families in crisis. We submitted the following as elements for review of this proposal:

1. Copy of our 2007-2008 Progressive Investment Report
   a. The 2007-2008 Progressive Investment Report documents the range of Building Human Asset Meetings that have been with federal, state, county, and local government officials to discuss ways to improve learning and support services to poor, minority, high-need, and special-need children, youths, and young adults through evidence-based reporting.

2. Copy of *Education and Science* (2009)
   a. The book *Education and Science* (2009) documents our study of why conduct concerns may relate to mental health when behavior issues and learning problems are not addressed in a timely manner; how and why learning problems may relate to academic performance in home, school, neighborhood, and workplace networks; and how and why conduct concerns may relate to mental health when behavior issues and learning problems are not addressed in a timely manner.

3. Copy of Community-Based Learning Certificate of Achievement Program

a. We work with children and parents in crisis to learn more about how children with traumatic injuries, behavior issues, learning problems, conduct concerns, and emotional disturbances live, learn, think, and respond in their home, school, neighborhood, and workplace networks while growing up hurt by these major life events; as they move through the K-12 school system. Based on 2000-2010 Evidence-Based Reports concerning the following:

   1. Harmful body or brain experiences
   2. Troubling action or responses to the experience of contact
   3. Poor experiences with interaction
   4. Lack of internal control
   5. The affect on action or responses to internal or external events

By developing a partnership with Save Our Youth, the Next Generations the County Department of Health and Human Services will be able to offer more options to

- high-need parents and parents working with children in crisis;
- foster care parents (guardians) and providers working with children in crisis;
- health care providers, nurses, and staff;
- educational program specialist, counselors, teachers, and school principals;
- mental health counselors, staff, and advisors;
- social workers, social service providers, and health and human service workers;
- corrections officers, staff, and probation officers; and
- learning, behavior, and mental health specialists.

The needs of children and families in crisis due to major life events in their home, school, neighborhood, and workplace networks are complex. Professionals who work with children and parents in crisis must possess a high level of expertise in the areas of human systems science problem solving methods, Human Systems Research, assessment in learning and support services, Progressive Investing models, and interaction techniques. To reduce the threat of cyclic decline, they must be able to integrate evidence-based practices that allow them to help children and parents from where they are—in the family, medical, educational, mental health, social work and correctional systems.

This is why our proposal can provide new and innovative options to improve the home, school, neighborhood, and workplace networks of children and families in crisis. Hence, the goals of the Building Human Asset Meeting process is to inform and instruct our leaders as we take these necessary steps to reduce family decline, school failure, delinquency, and poverty.

Human Systems Model 23

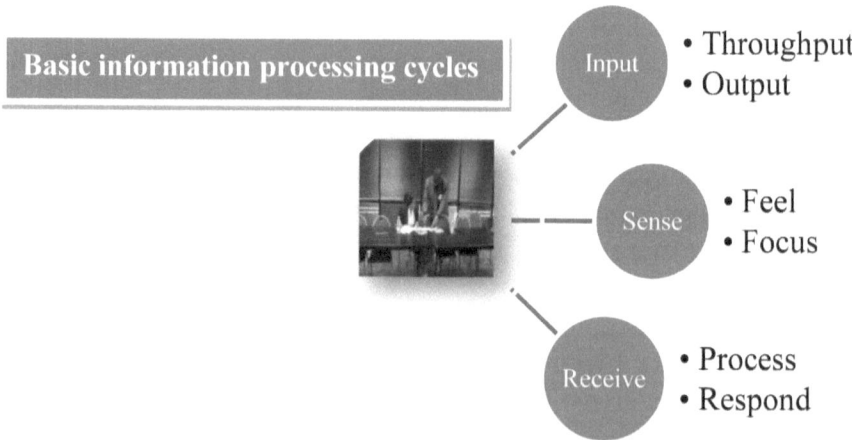

We informed Ms. Buckley: This phase of the PERCS project runs through June 25, 2010. When you think about it, Human Systems Research is the answer to reducing family decline, school failure, delinquency, and poverty for most poor, minority, and high-need parents, and children involved with poor home, school, neighborhood, and workplace networks. This is because, for the most part, these groups' major struggles involve learning how to use the land to live, learn, think, and respond to tension, stress, pressure, and competition.

Today, you have an opportunity to learn with Save Our Youth. We submitted the following program proposal as a workforce improvement product:

Progressive Investing Institute of Focused Learning In Partnership with Save Our Youth, the Next Generations

Parent Education and Resource Coordination Services Workshops
Community-Based Learning Certificate of Achievement

Overview

The needs of children and families in crisis due to major life events in their home, school, neighborhood, and workplace networks are complex. Professionals who work with children and parents in crisis must possess a high level of expertise in the areas of human systems science problem solving methods, human systems research, assessment in learning and support services, Progressive Investing models, and interaction techniques. To reduce the threat of cyclic decline, they must be able to integrate evidence-based practices that allow them to help children and parents from where they are—in the family, medical, educational, mental health, social work, and correctional systems.

Statewide children in crisis are experiencing one or more of the following threats to their growth, maturity, and development: family decline, school failure, delinquency, and poverty due to learning, behavior, and conduct problems that are not being screened, diagnosed, or assessed. As more and more children are being born into families already in crisis, the need for a human systems approach to preparing client-based systems for increased contact with them is very apparent. Our Parent Education and Resource Coordination Services (PERCS) workshops address this need by providing learning and support services to improve the home, school, neighborhood, and workplace networks of children and families in crisis.

In-depth

We work with children and parents in crisis to learn more about how children with traumatic injuries, behavior issues, learning problems, conduct concerns, and emotional disturbances live, learn, think, and respond in their home, school, neighborhood, and workplace networks while growing up hurt by these major life events as they move through the K-12 school system. Based on 2000-2010 Evidence-Based Reporting on the following:

1. Harmful body/brain experiences
2. Troubling action or responses to the experience of contact
3. Poor experiences with interaction
4. Lack of internal control
5. The affect on action or responses to internal or external events

Benefits

These workshops provide participants with the following:

- A core body of knowledge, skills, and practices in PERCS
- Parenting, dropout prevention, discipline and self-control, and job readiness strategies
- The skills and resources to screen and assess learning, behavior, and conduct disorders based on their influence in home, school, neighborhood, or workplace settings
- An opportunity to network with small learning communities
- An exciting opportunity for greater success working with children and families in crisis
- Community-Based Learning Certificate of Achievement

Audience

Workshops are geared toward parents, health care providers, educators, mental health professionals, social workers, and corrections professionals who work with children and parents in crisis:

- High-need parents and parents working with children in crisis
- Foster care parents (guardians) and providers working with children in crisis
- Health care providers, nurses, and staff
- Educational program specialist, counselors, teachers, and school principals
- Mental health counselors, staff, and advisors
- Social workers, social service providers, and health and human service workers
- Corrections officers, staff, and probation officers
- Learning, behavior, and mental health specialists

Structure

PERCS workshops are comprised of five courses, eight hours total, one day per month, over a twelve-month span. This is a small learning community made up of self, group, and environmental areas of focus. Participants should be committed to completing the project within the scheduled time frame. If participants miss a workshop, there is no guarantee it will be offered in the future. Each small learning community is made up of participants from the various fields noted above as an authentic Community-Based Learning Team.

Workshops

For workshop descriptions and schedule, contact Dr. **Christopher K. Slaton at (916) 363-6149** or saveouryouth@aol.com or visit the following:

> www.Drslatonprogressiveinvesting.com
> www.Saveouryouth.com
> www.Progressiveinvestmentgroup.org

To compete—live in a home, learn in a school, think in a neighborhood, and respond in a workplace—a child, youth, or young adult must be set up for contact with self, other people, and these environments. In the human system of man-made homes, schools, neighborhoods, and workplaces, to compete, a child must learn how to use a home to live, use a school to learn, use a neighborhood to think, and to use a workplace to respond.

Children that are not set up for contact with self, other people, and the environment do not interact well with uncertainty or change, because they lack an experienced sense of self, other people and the environment to help them live, learn, think, and respond to social demands. Troubles in the home, problems in the school, issues in the neighborhood, and concerns in the workplace lead to a poor self-concept that limits his or her ability and capacity to relate with other people who share these environments with them.

For instance, by our April 23, 2010, PERCS workshops, our parents were able to talk through feelings of hurt from their work experiences with children in crisis. We have learned from our home, school, and neighborhood visits with parents and their children in crisis. When a behavior issue goes unaddressed, it has an effect on learning, and that learning problems may be at the root of low tests scores and poor grades. Hence, an emotional disturbance is related to feelings of disappointment for the parent, the child, and the teacher. What this means is that the conduct, the pattern of carelessness, aggression, and sadness is related to the child's feelings of frustration.

We use our work with children and parents in crisis to learn more about how children with traumatic injuries, behavior issues, learning problems, conduct concerns, and emotional disturbances live, learn, think, and respond in their home, school, neighborhood, and workplace networks while growing up hurt by these major life events as they move through the K-12 school system. "This is always a tense and exciting process," a learning system (Checkland 1999, 219); to learn from the experience, when you use the systems method, but to focus your line of attack as a means of taking action to work the problem through.

Science is presented here as a system. This school of thought is concerned with the whole family and the children that emerge. Hence, the parent and child are seen as complex properties of a physical family system. In the next talk, we explain why our Parent Education and Resource Coordination Services workshops can be seen as an act of science. What follows, then, is the practice and reflection that emerged as a crisis intervention system we share and talk about how it feels to grow up hurt. This affect alone may impede a parent or child's skills to live, learn, think, and respond as a learning system.

Chapter 5

Parent Education and Resource Coordination Services (PERCS)

Following the processes we have outlined in Human Systems Research, ten years of collaborative action research (Sagor 1992) with parents, children, and their schools, led us to PERCS. For instance, using Building Human Asset Meetings with parents, children, teachers, school principals, and political leaders is an objective way to gather data to learn and understand the problem of family decline, school failure, delinquency, and poverty or lack of employability. This section bridges all the systemic components into a series of research-based interventions to improve learning and support services to all participants.

Hence, this book brings all the human learning and participant inquiry processes (Dewey 1922; Reason, P. 1994)—a comprehensive and research-based framework that has allowed us to test—how children and parents in crisis live, learn, think, and respond—how a child that has been hurt sense and receive data. We have explained the techniques for identifying and understanding these problems above. Now we turn to the learning and training process for a step-by-step paradigm shift through PERCS Workshops held in 2010.

Introduction to PERCS

Welcome to our Parent Education and Resource Coordination Services (**PERCS**) workshops. Save Our Youth the Next Generations (SOY) is a 501 (c) (3) nonprofit learning and training organization that has been delivering learning and support services to children and families in crisis since 1997. Working in partnership with SOY, we developed the Progressive Investing Institute of Focused Learning in 2000 to study child development through the application of action research, systems thinking, and human science.

Parents with children aged 0-25 (zero to twenty-five) gain access to advise on how to live each day to become more informed. In the process of contact between the parent and the advisor, the home, school, neighborhood, and workplace network is discussed and studied in relation to the parent, other people, and their child, youth, or young adult to learn about any relative learning, behavior, or conduct deficits. Building Human Asset Meetings may be used to set up certain contact and interaction with the people who share roles and functions in the parent's day-to-day home, school, and neighborhood, and workplace networks to plan, organize, and tests problem solving tools. Hence, parents may learn more about how the nurse, teacher, counselor, or psychologist and other professional support relates to their child's capacity and ability to move through the public school system to best benefit from a free and appropriate public school education.

A Systems-Based School of Thought for Crisis Intervention

Think about what it feels like to be human. To be human is the first goal in life to feel and develop important skills that will be needed to succeed as a person. Did you choose your parents or did your parents choose you? How did your life unfold from the embryonic stage to where you are right now? Was there a road map to follow? Who or what led you to be? Do you hear voices or do voices hear you? What does it feel like to know you have a human system to test, show off, and learn to use by doing things? If you are a person who lives in a state of crisis, you want answers. Here, we try to give you a few responses to these questions.

The Human System

- A child, youth, or young adult has to learn how to live with a special set of qualities that give him or her their potential to grow, mature, and develop as a physical, mental, social, emotional, environmental, sensory, language, and hence, human system.

The Mental System

- A child, youth, or young adult's human system accounts for their brain or body's capacity and ability to show a sense of self in relation to mechanics or academics as a mental system.

The Physical System

- A child, youth, or young adult's mental or learned responses to contact sets him or her apart as they can be observed to interact with self, other people, and the environment to live, learn, and think as a physical system.

The Environmental System

- A child, youth, or young adult has to understand how he or she lives, learns, thinks, and responds in a physical, mental, social, emotional, environmental, sensory, language, and human context that must be learned for the nature of these transactions to grow, mature, and develop the skills of their brain and body; capacities and abilities to live in the world; and to react as a green technology.

Crisis Intervention Strategy

We use Parent Education and Resource Coordination Services to help children and families in crisis gain access to the right family, medical, educational, mental health, social work, and corrections professional to reduce the risk of cyclic declines due to learning, behavior, and conduct disorders (as life threatening events) that were not diagnosed in accordance with a child, youth, or young adult's years of eligibility for a free and appropriate public school education. Parents are relieved when they can feel and see changes in the way they make contact with their child and, hence, the way their child makes contact with them. In this sense, the parent and the child are learning how to help the helper help them move feelings through the human system.

We want to know how the parent and child learn. When we use the word *human system*, we mean the body, senses, and brain. Hence, we work with the parent and child's body, senses, and brain to learn how they learn.

Human System

Human Systems Model 24

What is the human system?

1. The body is a physical object that can be experienced as an event.
2. The senses are the external objects and internal elements affected by body and brain activity.
3. The brain is a mental element in the body that has to be set up to live, learn, think, and respond to learning throughout life as a goal.

To learn about brain, body, and sense events, we look at how a person makes contact with self, other people, and their environments. This is what we mean when we say human systems research.

Human Systems Research

Human Systems Model 25

What is Human Systems Research?

1. The study of self is an event.
2. The study of the environment is a learning goal.

3. The study of the self in relation to the environment allows us to assess how learning (in crisis) affects other people.

As we have stated, we use words that emerge from our study of the brain, body, and senses like human system, and human systems research, to learn more about the functions of being human. Human systems domains are research-based terms in the sense that they have emerged from our work with the parent and child in relation to their home, school, neighborhood, and workplace interaction. Human systems domains are fields used to learn ways to work with the parent and child in crisis.

Human Systems Research Domains

Human Systems Model 26

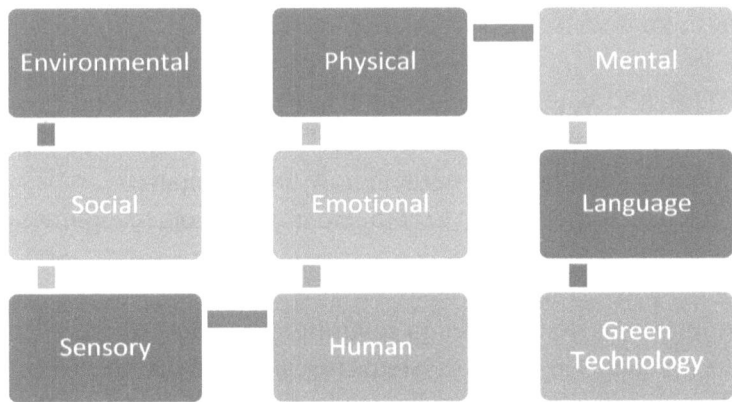

- Human systems research domains offer a structured format to learning how to work with a child, youth, or young adults body, senses, and brain; sense of self, other people, and the environment; and research-based themes to help the helper set up crisis intervention process cycles.

Competitive Levels of Organization

Human Systems Model 27

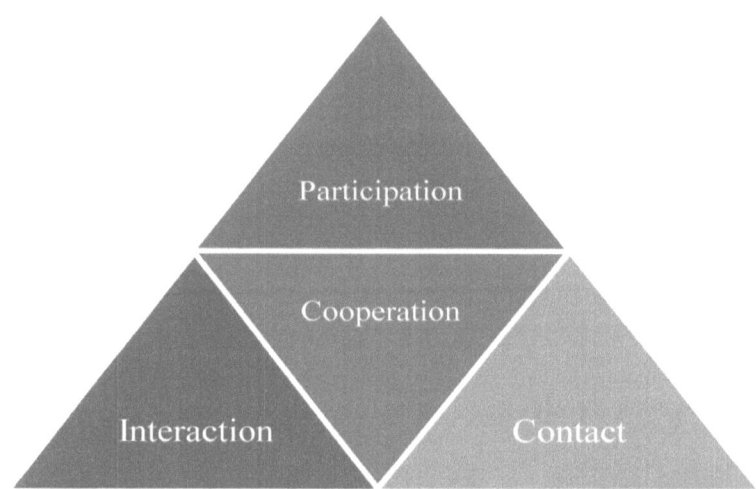

1. The child, youth, or young adult must learn through the experience of contact (plus interaction and cooperation) with self, other people, and the environment how to participate.
2. Each concept is an act to compete for attention, recognition, goals, success, and so on.

The human system moves in relation to contact, which as we have stated forms stages of human systems research and that the self interacts through process cycles to learn how to live through cooperative states to become more informed; this is an applied use of human systems domains to increase capacity and potential to participate through the use of talk and words to sustain the connection. These are competitive levels of organization through which self, other people, and their environments need to produce and sustain to reach sufficient stages of learning how to participate through acts to live, learn, think, and respond.

Competitive States of Taking Action

Human Systems Model 28

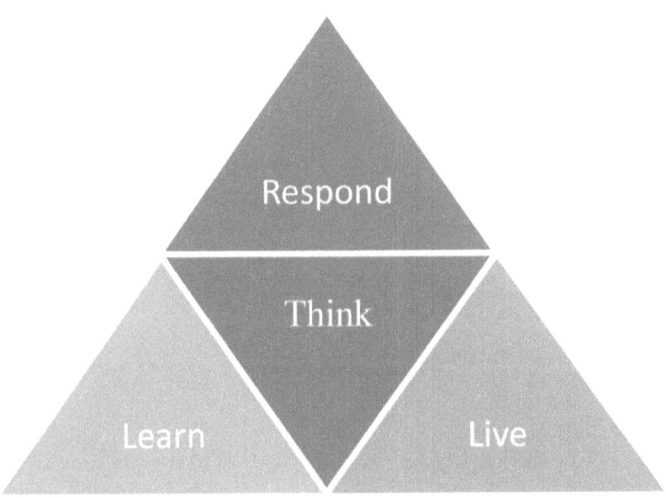

Taking Action:

1. The child, youth, or young adult must study ways to take action to improve how they live, learn, think, and respond to self, other people, and the environment.
2. In each process variable, the person is learning how to compete to live in a home, how to compete to learn in a school, how to compete to think in a neighborhood, and how to compete to respond in a workplace.

Competitive Stages of Experience

Human Systems Model 29

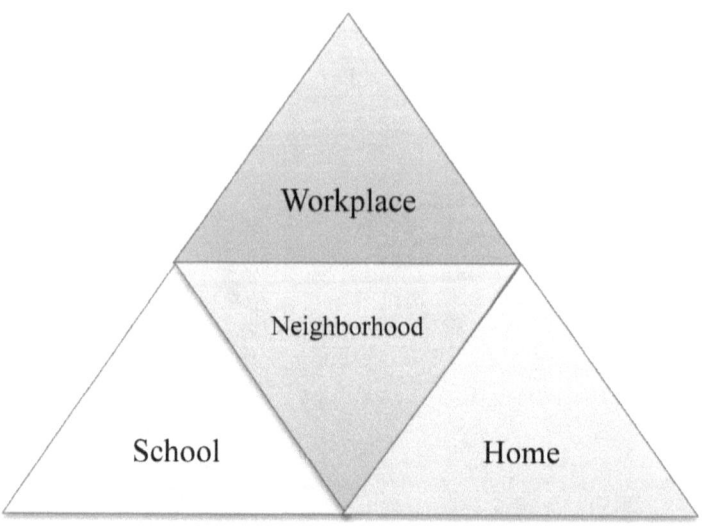

1. The child, youth, or young adult must learn how to experience the act to live in a home, to learn in a school, to think in a neighborhood, and to respond in a workplace through the study of self, other people, and the environment.
2. Each place is a setting where competitive relationships emerge in connection with land use, becoming informed, ownership of property, and wealth building.

These competitive stages of experience are hurt or aided by the way life is at home, in school, in the neighborhood, and in relation to the workplace as separate, but connected social places where levels of contact, interaction, cooperation, and participation do, or do not, organize.

Competitive Systems to Explore

Human Systems Model 30

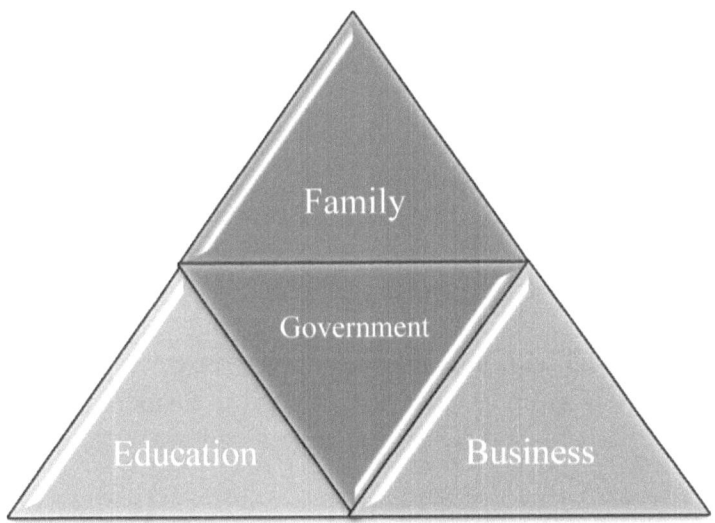

1. The child, youth, and young adult must be led to explore family, education, government, and business affairs in the search for a higher sense of self, other people, and the environment to learn and come to understand the American economy.
2. Each practice is a competitive stage in the growth of a family as a business, a child as an investment, and the community as a place to learn by doing things to improve the quality of life.

To a family that is in crisis, these are stages in their economic growth and in their organization of human skills. This is determined by how well the brain, body, and senses are set up to use the land and to explore human affairs to compete for ways to live, learn, think, and respond through sufficient levels of participation.

Community-Based Learning Academies

We are here to help. Parents in crisis, because of family decline, school failure, delinquency, poverty or lack of employable skill sets, need to learn how to use their home, school, neighborhood, and workplace network.

These are the four main components of Community-Based Learning:

1. Home-Based Learning (0-5)
   — Parents and children that have been hurt by major life events, learn how to live together with a focus on contact, interaction, and cooperation inside the home.

   1.2 Dealing with Social Issues in the Home
      — Attachment Disorder (related to abandonment, neglect, abuse, and separation from loved one)

2. School-Based Learning (6-10)
   — Teachers and children that have been hurt by major life events learn how to learn together in school, with a focus on contact, interaction, and cooperation inside the classroom.

   2.1 Dealing with Problems in the School
      — Academic Skills Disorder (related to learning, behavior, or conduct)

3. Neighborhood-Based Learning (11-15)
   — Parents, teachers, and youths that have been hurt by major life events learn how to think through neighborhood projects, with a focus on how they make contact, interact, and cooperate to participate in learning how to live, learn, and think together.

   3.1 Dealing with Concerns in the Neighborhood
      — Delinquency (related to isolated sense of self, resisted sense of others, and rejected sense of the environment).

4. Workplace-Based Learning (16-25)
   — Young adults that have been hurt by major life events learn to how focus their brain and body through service learning, with the workplace as the goal to aim the way they respond to live in a home, learn in a school, and think in a neighborhood.

   4.1 Dealing with Specific Events in the Workplace

      — Poverty (related to social issues in the home, problems in school, and concerns in the neighborhood).

Move through the Crisis

PERCS is designed to prepare parents, teachers, students, school principals, and other social service agents for contact to improve the delivery of learning and support services to parents and children in crisis. Recent meetings with school and political officials have led us to understand why children in crisis are not being moved through the public school system with greater amounts of care. When a parent and child are receiving services through the use of multiple human service agencies, this is the best way to organize a system, process, and strategy.

Building human assets through working with parents and children in crisis is based on basic scientific methods. In essence, we propose that scientific methods be used in connecting the home to school, the school to the neighborhood, and the neighborhood to home and school as a human network. When parents and children practice learning how to live with self, other people, and their environments, they become more effective resources. Preventive practices change how parents and children receive the intervention, when they are asked to participate in the enforcement of learning goals that have been identified and explained to them.

Progressive Investing Systems connect the process cycles from one experience to the next in an intricate send-and-receive loop. We are working on ways to improve the delivery of learning and support services to children and parents in crisis. Process cycles change behavior since each act intersects with the flow of information, knowledge, experience, and reflection. Hence, these are participatory systems that allow the person to practice learning through the flow of action. We do this work to build a parent and child's skill to respond to contact in the crisis situation and interact to change the behavior or effect.

Closing Reflections

How do you feel things?
What have you learned?
Why is the experience of feelings vital?

Welcome to the Crisis Intervention Conference for Males

I want to thank Blake Johnson and Devon White for spending time with me as we developed our current conference. Both Blake and Devon are high-achieving males that have enjoyed a great amount of success in school. It was truly a pleasure to sit and chat with them about the plight of our males in crisis.

We have held many Building Human Asset Meetings with males since 2000 with a focus on ways to improve the delivery of learning and support services. For instance, the first step we took was to collect authentic data to build up a crisis intervention system (CIS) for males. We have studied how males make certain decisions in relation to family, education, government, and business affairs to define the skills and patterns of behavior that need to be set up for males in crisis to experience.

By taking these steps to learn how to help, we are in a better position to offer problem-solving tools to teach the skills and behaviors that need to be enhanced for males to compete more effectively in the twenty-first century. Hence, we think that it is vital for males in crisis and males in general to learn how to receive information, process knowledge, respond to the experience of experience, and lead through intelligent reflection.

The requirement for these skills are (1) the ability to use clear talk, (2) the ability to learn through change, (3) the ability to feel how it feels to be a friend to a female, and (4) the ability to move through the initial experience of hurt. As we move forward, the data we have collected for this conference will need to be assessed. This means, of course, you will need to feel for this experience to unfold through your analysis of its value.

The following is a sample press release:

One Day Crisis Intervention Conference Press Release

The Progressive Investing Institute of Focused Learning ™ 2003

- School Code: 3407311

What: Crisis Intervention Conferences 301 and 302

When: Friday, March 26, 2010

Where: Fairfield Inn & Suites

   10745 Gold River Drive
   Rancho Cordova, CA 95670

Crisis Management Sessions

1. 8:00-9:30 AM (Crisis Intervention Conference 301: for Males ages 9-15)
2. 1:00-2:30 PM (Crisis Intervention Conference 302: for Males ages 16-25)

To Register: Contact Dr. Christopher K. Slaton

- Email: slaton@softcom.net
- Tel: 916-955-1368
- Web: www.Progressiveinvestmentgroup.org
- Web: www.Drslatonprogressiveinvesting.com

Who: Parent(s), fathers, brothers, sons, daughters, aunts, uncles, guardians, and other helpers

Why: According to our evidence-based research, many child, youth, or young adult males are in crisis due to major life events that affect how boys and men live, learn, think, and respond to self and other people.

How: We will discuss mild to moderate learning, behavior, and conduct problems with a focus on ways to work with males with severe learning, behavior, and conduct problems that affect how they live in a home, learn in a school, think in a neighborhood, and respond in a workplace.

Key Feature: Human Systems Research is the most influential aspect of building human assets, because it relies on the study of self, other people, and the environment to learn how to improve home, school, neighborhood, and workplace networks. Since 2000, we have committed a majority of our resources to research and development of new approaches that improve the lives of children and families in crisis.

Responses: Parent reflection is a powerful way to assess the work we do with children and families in crisis. We close each workshop with reflection to learn firsthand what you feel has been experienced. We help you improve your responses to contact through interaction.

Opening Talk

Crisis intervention conferences for males are human learning centers we use to help parents and children who accept our invitation to participate.

1997-2010

Jason and Garret are brothers helping us talk about ways to move through crisis.

> We work to improve learning and support services to poor, minority, high-need, and special-need children, youths, and young adults.

When you are in crisis, you move through it to learn from the experience of your own action. Action refers to the way you work through changes in how you feel the experience in social places where people can see you, listen to you, and feel you too.

Males in crisis with moderate to severe learning, behavior, or conduct problems enter the Progressive Investing Institute of Focused Learning. Action research, systems thinking, and human development is fused as a method to move the male in crisis to learn how to participate, organize the senses, and work through emotion. Each year, new Progressive Investing models are being built up or enhanced. This improves our assessment in learning and support services as a human systems science problem-solving method.

1. We study how and why behavior issues may relate to life inside the home.
2. We study how and why learning problems may relate to academic performance in home, school, neighborhood, and workplace networks.
3. We study how and why conduct concerns may relate to mental health when behavior issues and learning problems are not addressed in a timely manner.

Sensory Organization

Males are asked to talk to their brain through essay writing. Males learn from the outside to the inside, how to feel for brain, body, and sense connections. In other words, he will experience the senses through contact, the way the body interacts, and how the brain cooperates to live.

Sensory Problem Solving Essays (Writing to the Brain)

- Who am I?
- What am I?
- Why am I here?
- Where am I at my best?
- When is the best time to be me?
- How do I live, learn, think, and respond to contact?

Information, Knowledge, Experience, Reflection:

To become informed, to acquire knowledge, to feel experience, and to practice reflection are basic goals that move through the events of self-analysis to signs of an affect on brain, body, and sense connections. Males practice learning, in the sense of how to feel things to learn from the contact, ways to receive data as signs of care.

1. **CMS 1: Growing Up Hurt by Major Life Events**
   - We will discuss how growing up hurt is a threat to your ability to live and respond to contact. Hence, a threat may be any event at home, at school, in a neighborhood, or in a workplace that reduces your capacity to learn and think things through to interact. This CMS card is for purposes of education and research only.

2. **CMS 2: Growing Up Hurt by Major Life Events**
   - We will discuss how the way you enter contact has an effect on your desire to interact to participate. To participate in home, school, neighborhood, and workplace events you must be willing to cooperate. Hence, you must be able to physically display your ability to live with self, respond to other people, and learn to mentally display your capacity to use these events to think things through. This CMS card is for purposes of education and research only.

3. **CMS 3: Growing Up Hurt by Major Life Events**
   - We will discuss the lack of the ability or will to think things through is a crisis in any event that requires the capacity to learn new things. This is because to live in a home, to learn in a school, to think in a neighborhood, and to respond in a workplace requires an ability to learn new things. Hence, these skills are needed to deal with personal, academic, social, and occupational issues, problems, concerns, and goals that have an effect on your learning, behavior, and conduct. This CMS card is for purposes of education and research only.

For a child, youth, or young adult that has been hurt by major life events, a key problem may be internal control of the body, brain, and senses to pursue acts and goals of the group (family) instead of the self (individual). This may mean that the authority structure is resisted, and the rewards for conformity are at odds with the individual's ability to perform, because they cannot appreciate the penalty. This is why a crisis management system can be successful, if every act and goal is aimed at the child, youth, or young adult's brain in a focused effort to organize the brain, body and the senses.

Closing Reflections

How do you learn new things?
What have you felt?
Why do we say feelings of thought are more vital?

The Screening and Assessment Process

We work with children and parents in crisis to learn more about how children with traumatic injuries, behavior issues, learning problems, conduct concerns, and emotional disturbances live, learn, think, and respond in their home, school, neighborhood, and workplace networks while growing up hurt by these major life events as they move through the K-12 school system.

Information Processing Cycles

Information processing cycles may influence how a person is able to sense, feel, and focus their body and brain.

Information is gathered through:

Human Systems Model 31

- *Physical*
- *Mental*
- *Social*
- *Environmental*
- *Emotional*

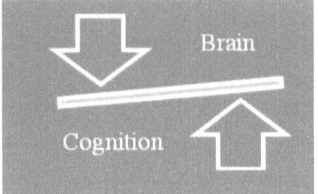

Action
Contact
Interaction
Cooperation
Participation

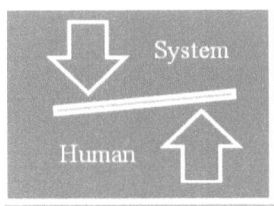

Event

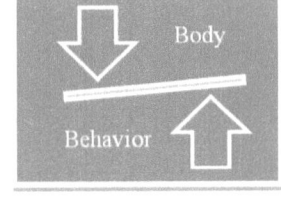

Effect

Social and Academic Learning

We use the word *social* to refer to observable events that interact with the senses, and the word *academic* to refer to fixed events that are theoretical and need to be set up for the brain to receive. In other words, the social act to sense is physical and the academic act to receive is abstract.

Human Systems Model 32

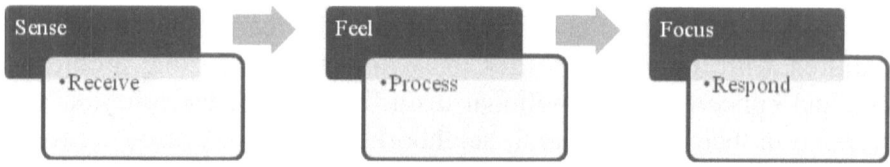

1. For social learning, the body has to be set up.
2. For academic learning, the brain has to be set up.
3. For social and academic learning, the senses have to be set up.

Discipline and Self-Control

Recall that discipline is physical and that self-control is mental.

Human Systems Model 33

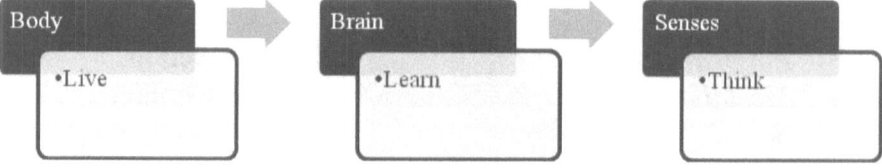

1. The body is a sign of physical discipline.
2. The brain is a sign of mental self-control.
3. The senses form signs of social and academic awareness through discipline and self-control.

This means the body, brain, and senses are designed to move through contact as a system through the use of reception to act and to interact.

Intrinsic Value of Information Processing Cycles

Human Systems Model 34

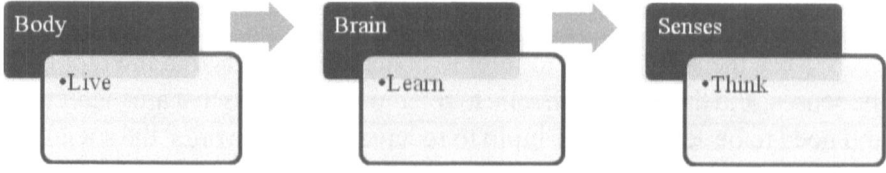

- The child, youth, or young adult gains common sense from the act to learn how the body, brain, and senses are set up to show signs of care.

Intrinsic Value of Information Processing Cycles

Human Systems Model 35

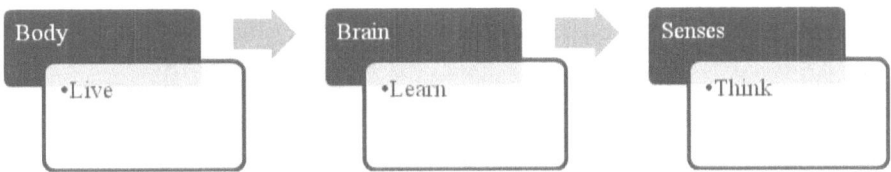

- The child, youth, or young adult builds up the courage to live through acts of discipline and self-control that govern the body and brain's behavior and learning.

Intrinsic Value of Information Processing Cycles

Human Systems Model 36

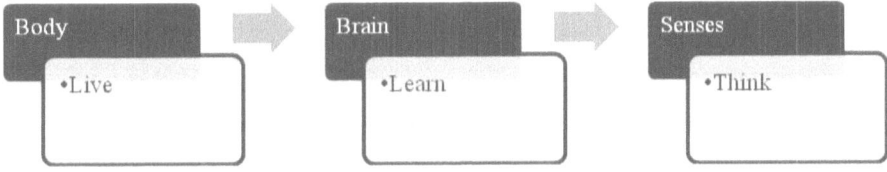

- The child, youth, or young adult builds up a sense of objectivity as knowledge on how to live results in discipline and self control sustained through the governing of needs and desires that set the human system up to feel the things one cannot change and to focus the senses to accept it (in fairness).

Thank You!

This is what we think science and education is all about. It is about the serious nature of how information is processed (Wilson 1991). The facilitation and data-collection processes we use reflect on the intervention practices noted in No Child Left Behind Act of 2001 and the Individuals with Disabilities Education Improvement Act (U.S. Department of

Education 2003; Chidsey and Steege 2005; Pierangelo 2003). Building Human Asset Meetings to improve learning and support services to poor, minority, high-need and special need children, youths, and young adults locked in poor home, school, neighborhood, and workplace networks were held with Lana Fraser, assistant deputy director at the State Department of Rehabilitation (2007); Pat Ainsworth, California Department of Education (2007); Larry Buchanan, superintendent, Del Paso Heights Elementary School District (2007); Steven Ladd, superintendent, Elk Grove Unified School District (2007); and Magdalena Mejia, superintendent, Sacramento City Unified School District (2007).

Closing Reflections

How do you feel?
What did you learn?
Why are you here?

Welcome to Our
Screening and Assessment Process for Working with Parents and Children 0-12

By advocating for the early screening and assessment of children, basic facts about how they entered life will become more known through observations of how they sense and receive data. Hence, children that come to school lacking the ability to receive, organize, and express their feelings are in crisis. This is because, in order to move through contact, the child has to interact with the way it makes him or her feel. In this sense, a child has to process the experience of contact to organize and focus the flow of feelings to brain.

Preparing the body and brain for contact with self, other people, and environments

Human Systems Model 37
1. Language Skills
2. Cognitive Skills
3. Physical Skills
4. Mental Skills
5. Social Skills
6. Environment Skills
7. Emotional Skills
8. Sensory Skills

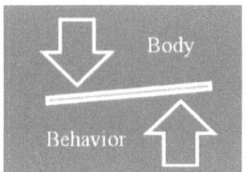 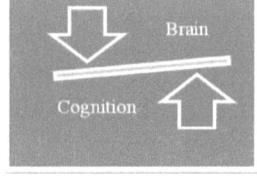

9. Language Skills
10. Cognitive Skills
11. Physical Skills
12. Mental Skills
13. Social Skills
14. Environment Skills
15. Emotional Skills
16. Sensory Skills

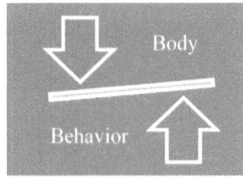 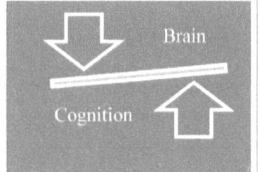

Preparing a child's brain for contact prepares the child's body to sense, feel, and focus how they learn to live with people.

Human Systems Model 38

Preparation Process for Life inside the womb:

Language Skills
1. Talking
2. Reading
3. Writing

Cognitive Skills
1. Learning how to talk to the brain
2. Learning how to read to the brain
3. Learning how to write to the brain

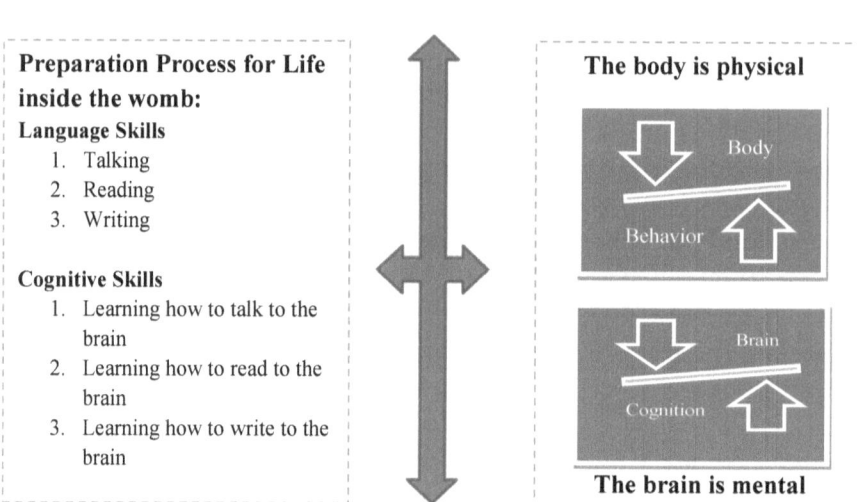

The body is physical

The brain is mental

During the preconception and embryonic stages, a parent has to set their brain and body up to care for the life of a child and to receive a child's contact.

Human Systems Model 39

Preparation Process for Life outside the womb:

Physical Skills
1. Patience
2. Sacrifice
3. Discipline

Mental Skills
1. Learning how to live with a calm state of mind
2. Learning how to survive with a sense of character in mind
3. Learning how to feel for a sense of self-control

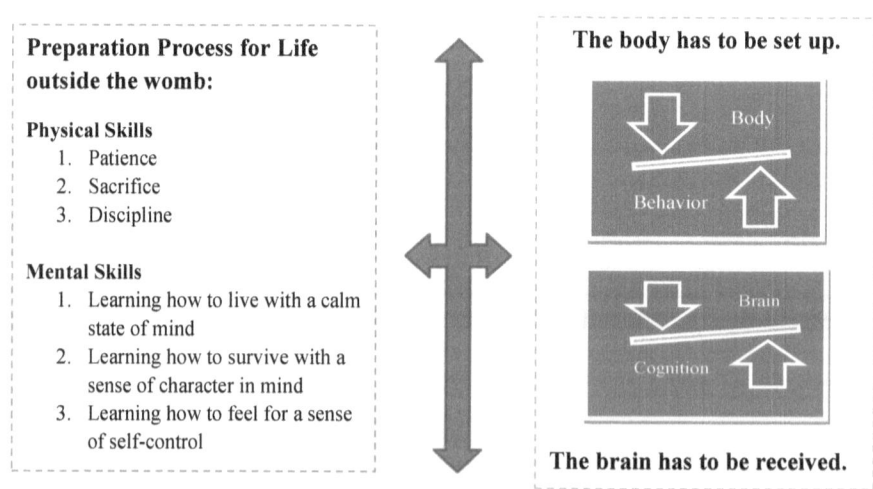

The body has to be set up.

The brain has to be received.

Between infancy and age nine, a parent has to set up a child's brain and body to sense, feel, and focus how he or she receives contact.

Human Systems Model 40

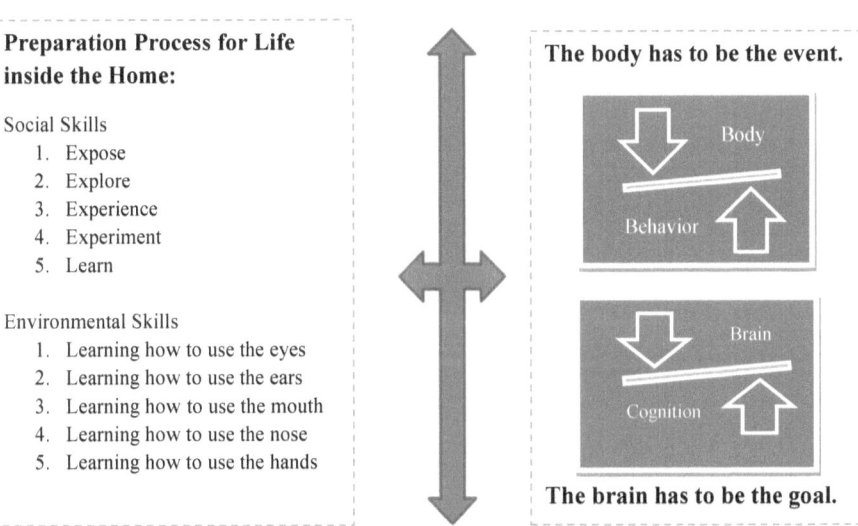

Between the ages of ten and twelve, a parent has to set up a child's brain and body to sense, feel, and focus how he or she receives contact.

Human Systems Model 41

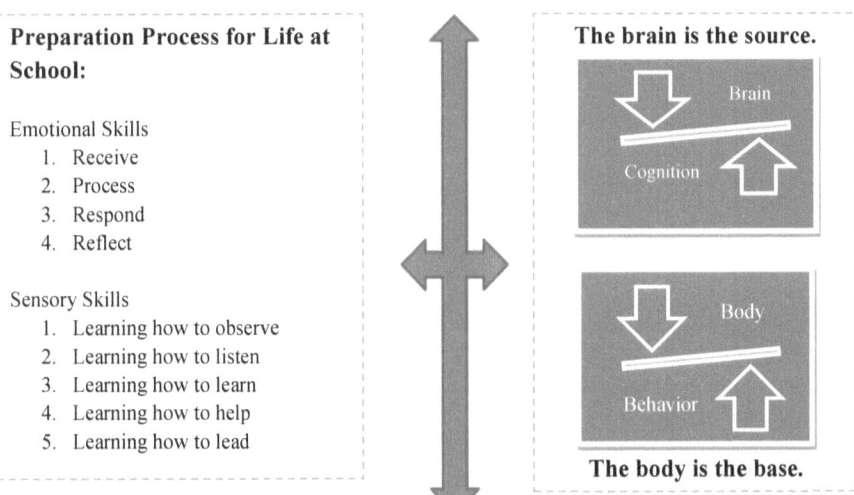

Thank You!

Kevin Johnson, mayor of the city of Sacramento wrote, "Thank you for contacting my office. I share your concerns for the education of our youth in Sacramento" (November 9, 2009). Lyn Corbett, director of the Office of Youth Development for the city of Sacramento wrote, "It was a pleasure meeting you. I look forward to working with you in the future" (March 1, 2010). "From this discussion, it should be apparent that family well-being as a value is different from the issues in which it may be at stake, abortion and unwed parenthood being examples. Whereas topical issues are matters of contention and controversy that come to the fore and then recede in importance over time, values such as family well-being tend to persist, cutting across topical issues" (Zimmerman 1995, 9).

Closing Reflections

What is a talking system?
How do you show signs of care?
When is the best time to be you?

Welcome to Our
Screening and Assessment Process for
Working with Males in Crisis CNC Workshop
Working with Parents and Children 10-25

At this stage, a child that is growing up hurt will show a lack of ability to read, write, listen, and talk through the use of academics to interact. This means that their senses are cut off from the experience of contact, which then forms a resistance to learning because of the amount of effort it takes to focus and sustain signs of discipline. Once a child falls behind, in the sense of moving through the contact to interact and cooperate as signs of a will to participate, they have rejected their need to feel and focus the brain and body to receive. Hence, he or she will lack signs of a controlled state.

The Experience of Emotions

Human Systems Model 42

1. Language Expressions
2. Cognitive Patterns
3. Physical Discipline
4. Mental Self-Control
5. Social Awareness
6. Environmental Care
7. Emotional Body
8. Sensory Organization

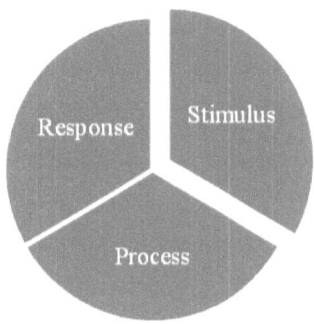

Language Expressions

Human Systems Model 43)

1. Noises
2. Sounds
3. Signs
4. Symbols
5. Proof of Interaction

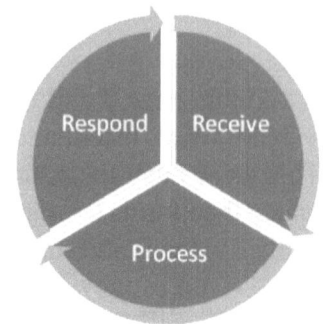

Cognitive Pattern

Human Systems Model 44

1. Sense
2. Feel
3. Focus
4. Proof of Ability

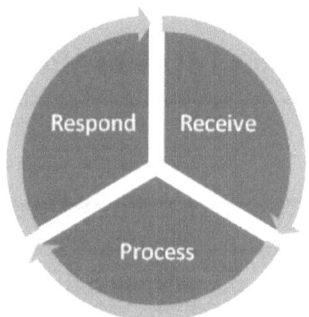

Human Systems Model 45)

1. Identity
2. State of Body
3. Sense of Meaning
4. Proof of Control

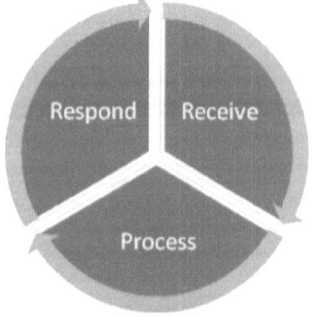

Mental Control

Human Systems Model 46

1. Action
2. State of Mind
3. Reasoning
4. Proof of Discipline

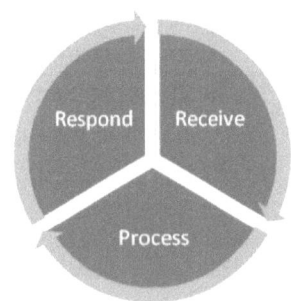

Human Systems Model 47

1. Sense of Self
2. Sense of Other People
3. Sense of the Environment
4. Proof of Contact

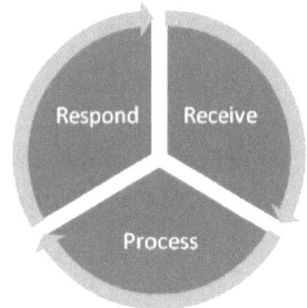

Environmental Care

Human Systems Model 48

1. Human System
2. Learning how to Live
3. Social and Academic Action
4. Helping the Helper
5. Proof of Cooperation

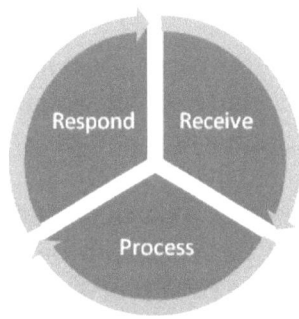

Emotional Body

Human Systems Model 49

1. Human Learning
2. Response to Sensing Things
3. Reaction to Feeling Things
4. Proof of Capacity

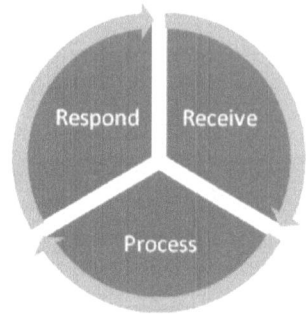

Sensory Organization

Human Systems Model 50

1. The Event
2. The Aim of the Goal
3. The Act of Moving through the Affect
4. The Experience of the Effect
5. Proof of the Participation

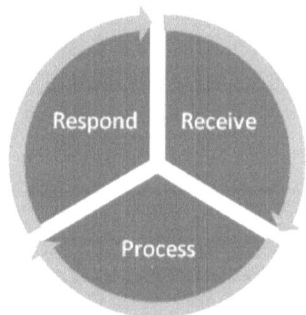

Thank You!

"As parents, adults, citizens and leaders we must examine ourselves regularly to determine whether we are contributing to the crisis our children face or to the solutions they urgently need" (Children's Defense Fund 2007, 2). Dr. Shelton Duruisseau wrote, "Thanks for not giving up on my participation in the critical work that you are doing. I will put the workshop on my calendar. If I am not able to attend, I will have a representative participate to determine if there could be mutual collaboration on behalf of these young people" (December 14, 2009). We wrote to Rancho Cordova School Leaders: "The parent education and resource coordination services workshops allow parents in attendance to think about how he or she sets his or her body up for contact with their child. This means that the parent is thinking about the language and tone of voice he or she uses when interacting with their child's brain" (April 26, 2010). "So many poor babies in rich America enter life with multiple strikes already against them: poor prenatal care, low birth weight, born to a teen, poor, and poorly educated single mother and absent father. At crucial points in their development, from birth through adulthood, more risks and disadvantages cumulate and converge that make a successful transition to productive adulthood significantly less likely" (CDF 2007, 3).

Closing Reflections

>Why are we here to help?
>What are you prepared to do to help?
>Where is the best place to start helping?

Welcome to Our
Screening and Assessment Process for
Working with Parents and Youths/Young Adults 13-18

This is the stage of social drift. A child that is growing up hurt by issues in the home, problems in school, and concerns in their neighborhood search for a place to fit in and feel for a sense of belonging. The tension of not knowing how to feel success, the stress of not being able to feel success, and the pressure of not being able to successfully feel for self, other people, and their environments push youths to behave in ways that that cause us to question their identity.

The human system begins to reorganize the way the brain, body, and senses are used to feel things. This means that the youth is being set up for the experience of becoming an adult. Youths have to calm down to move through feelings of contact that may upset the human system.

Human Systems Model 51

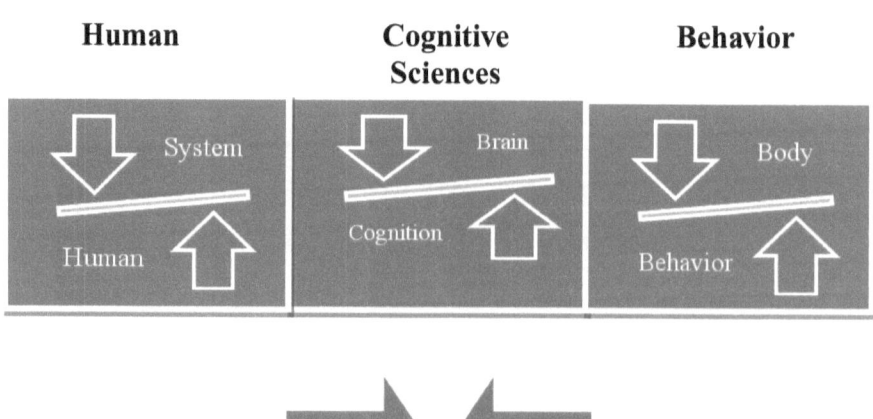

Language	System
Cognitive	System
Physical	System
Mental	System
Social	System
Environmental	System

Emotional   **System**
Sensory   **System**

Human systems Model 52

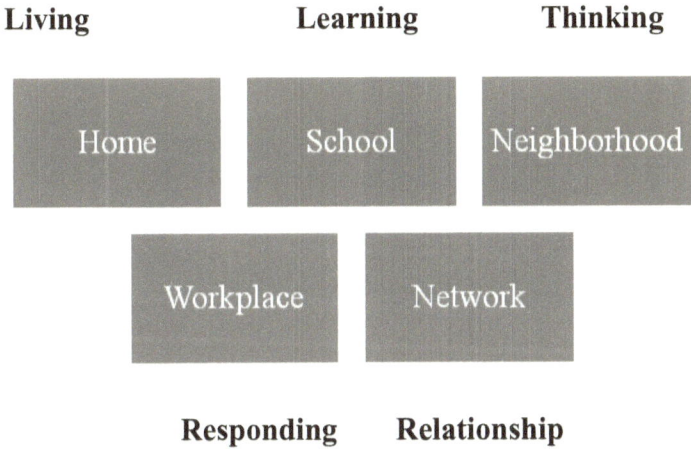

The human system has to be set up for living in a home, learning in a school, thinking in a neighborhood, responding in a workplace, and networking in a relationship. Youths and young adults need to live through contact in the home to learn how to interact in the school, which means to think and cooperate in the neighborhood, and respond to participate in the workplace. In each network, the relationship between self, other people, and their environments are being realized.

1. Youths and young adults *practice participating* from where they are.
2. Youths in crisis may not be ready to *practice participating* from where you or I may want them to be.

Human Systems Model 53

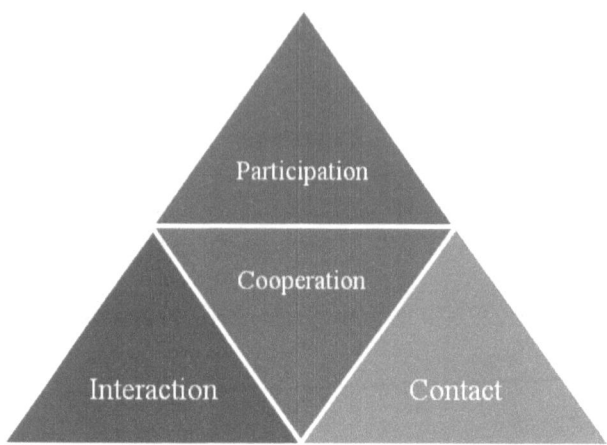

1. They move through these stages of participating in relation to the middle and high school grades they are experiencing.
2. They are experiencing feelings about self as an event, feelings about other people as an affect, and feelings about the environment as an effect.

Human System Model 54

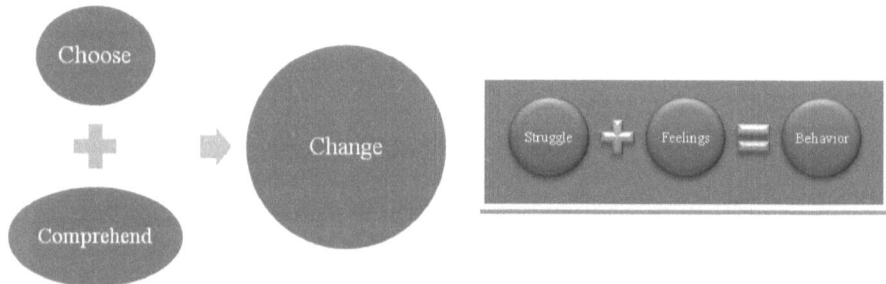

1. They systematically pull together ideas about the experience of home in relation to school and home and school affairs in relation to their neighborhoods.
2. They struggle to make sense of how it feels to live in relation to learning, learn in relation to behavior, and the need to focus in relation to thinking.

Human Systems Model 55

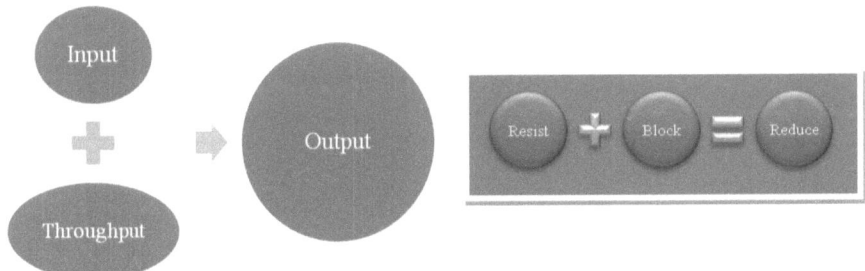

1. They understand behavior as an applied approach to control and manage their interactions with self, other people, and the environment.
2. They choose to resist the need to live to comprehend the need to learn how not to block the need to think, to reduce the threat of being changed, from the outside in.

Human Systems Model 56

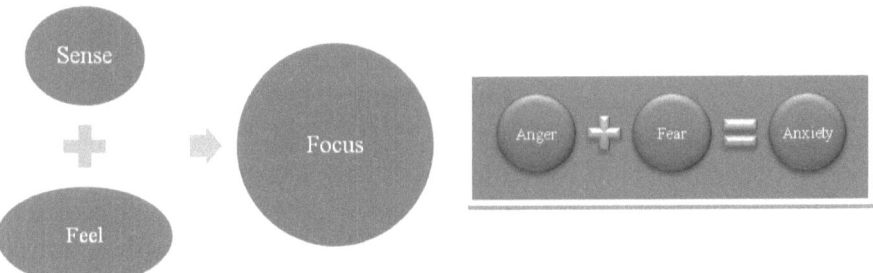

1. They use their emotions to resists the relevant values and goals to be achieved through their participation.
2. They use their sense of anger to feel fear, to release a focused act of anxiety, to reduce the need to care.

Human Systems Model 57

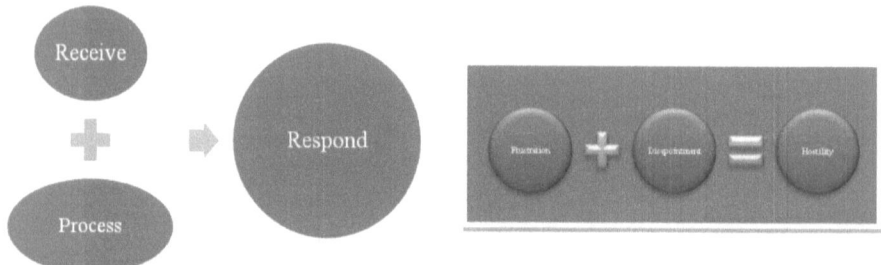

1. They use specific techniques and skills available (other people) to reject social and academic systems that seek to manage and control their behaviors.
2. They use the skills they have learned from parents and peers to receive the frustration they feel, to process the hostility they have experienced, to respond to the disappointment they cannot focus, and to reduce their inner sense of pain and sadness with self.

Their brains are either connected or disconnected from the social and academic systems they need to enter the workplace as young adults.

Thank You!

"In Human Systems Research, the conclusion of an analysis seeks to explain its connection to the ongoing study of how humans with problems live, learn, think, and respond. Hence, this has been a formative analysis of how children that have been hurt by major life events in the home, school, neighborhood, and workplace networks need to be set up. In each of our previous discussions, we have tried to focus on the needs of the whole child rather than on simple and isolated parts. This is because we believe in learning what a human being is, in the sense of what a human can do. This has been an act to sense, feel, and focus our analysis of the human system on synthesis" (Slaton 2009, 189). On March 26, 2010, we were requested to speak about our work with parents and children in crisis before approximately 125 guests at the "Empowering Youth for a Successful Future" scholarship dinner. We talked about ways to empower youth as green technologies and as wireless human systems. In twenty minutes, we explained how the body lives, the brain learns, the human system thinks, and how human systems research responds from the perspective of family leadership and child development.

Closing Reflections

How do you feel about this work?
What did you experience during our talk?
Where is the best place to be you?

Welcome to Our May 22, 2010 PERCS Workshop

Human Learning Goals Influence the Nature of Living, Learning, Thinking, and Responding as a Human System

Human learning goals identify the aim of brain, body, and sense events. How a child, youth, or young adult learns to feel contact will influence their natural acts to live, learn, think, and respond to it as a human system. Hence, the human system is set up to live, learn, think, and respond through the learning of goals that unify the tasks and meanings of brain, body, and sense events.

What do we mean by being human?

- Being human is an expression of aiming to live to become more informed through the event. Learning how is the desired goal state. The effect of aming is measured by the affect of the event and the goal state.

Progressive Investing Model 13

- The act to think is a cognitive construct between the event to receive contact, the goal to respond to the interaction, and affect, which comes from the act to process the event and goal to cooperate.

What do we mean by human learning goals?

- Human learning goals are created in the sense that they are responses to human activity.

Human Systems Model 58

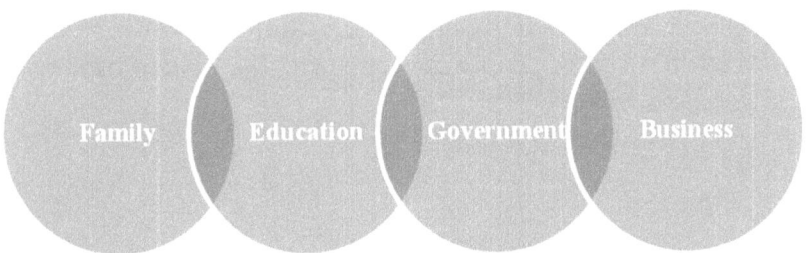

- The system of family, education, government, and business affairs are expressions of human activities designed to focus the way you choose to live, learn, think, and respond to self, other people, and the environment.

The Self as the Event

- Learning from the act to live: The purpose is to send internal feed-forward or to receive external feedback to learn and to feel the value of the act and to measure a sense of self.

Progressive Investing Model 14

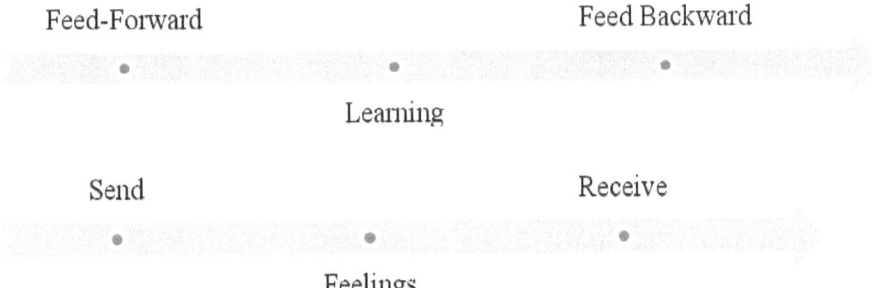

Other People as the Goal

- The process of energy: Human energy systems receive, process, and respond to other people as events within a field of experience.

Progressive Investing Model 15

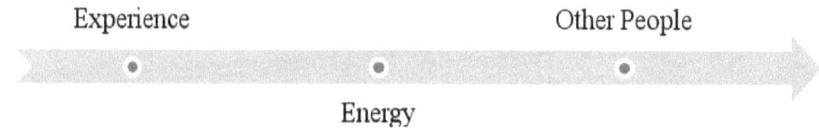

- The experience of other people affects the flow of energy through receive, process, and respond cycles.

The Environment as the Affect

- The field of experience: The home, school, neighborhood, and workplace network is affected by the contact, relationships, and the responses of self and other people.

Progressive Investing Model 16

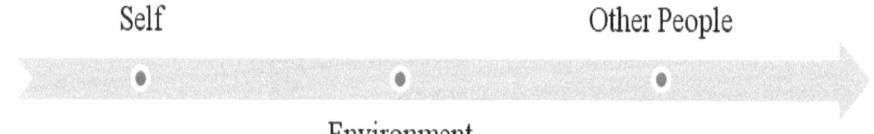

- The event to send a sense of self forward and the goal to receive experience from other people has an affect on the flow of energy through the environment.

Human Systems Model 59

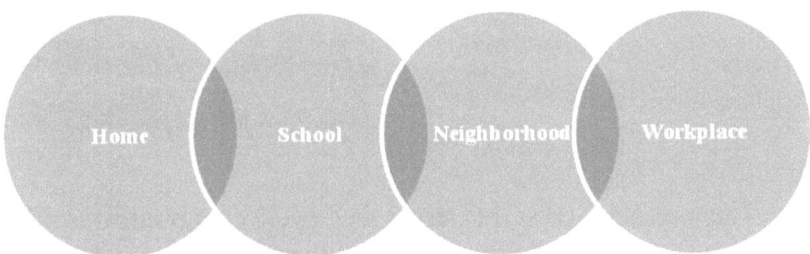

Goal Directed Learning

- Human, cognitive, and behavioral responses to self, in relation to other people and their contact with the environment, sets up the field of experiences in which learning goals are directed.

Human Systems Model 60

- The event of being a human system relates to the goal state as a behavioral response, which has an affect on how cognitive structures are set up.

Progressive Investing Model 17

Communication Systems

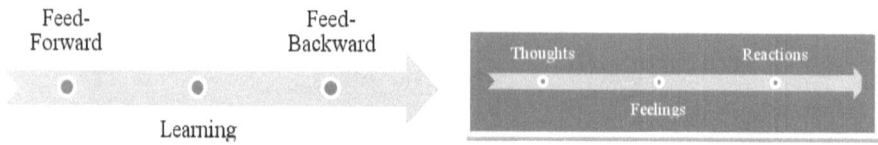

- The internal feed and external feed cycle communicates through a sense of self and other people to think and react to feelings.

Progressive Investing Model 18

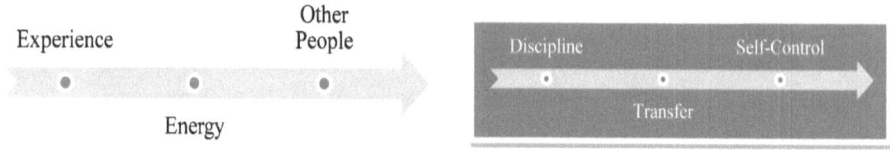

- Discipline is physical (human body) and self-control is mental (human brain), which manages the flow of energy to adjust to emotional conditions and to transfer data (human senses).

Goal Directed Action

Goal-directed action is energy and feelings in the process of becoming aimed.

Human Systems Model 61

- The act to achieve (family) leads to the objective (education), which leads to the purpose (government), which leads to the mission (business) to live, learn, think, and respond to

1. Live for family   Contact   In the Home
2. Learn for education   Interaction   In the School
3. Think for government   Cooperation   In the Neighborhood
4. Respond for business   Participation   In the Workplace

Open Human Systems

Open human systems sense and receive data in the aim to release a processed response.

Progressive Investing Model 19

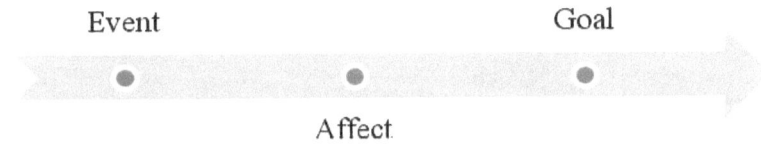

- When the child is open to contact, they grow from the experience of the event, they mature with other people as the goal, and they develop the energy to move through contact, to interact, to cooperate, and to participate as an affect on the environment.

Formal Human Systems Model

- Human learning goals make up a formal human systems model for building up the child's sense of self, other people, and the environment (Slaton 2009).

Human Systems Model 62

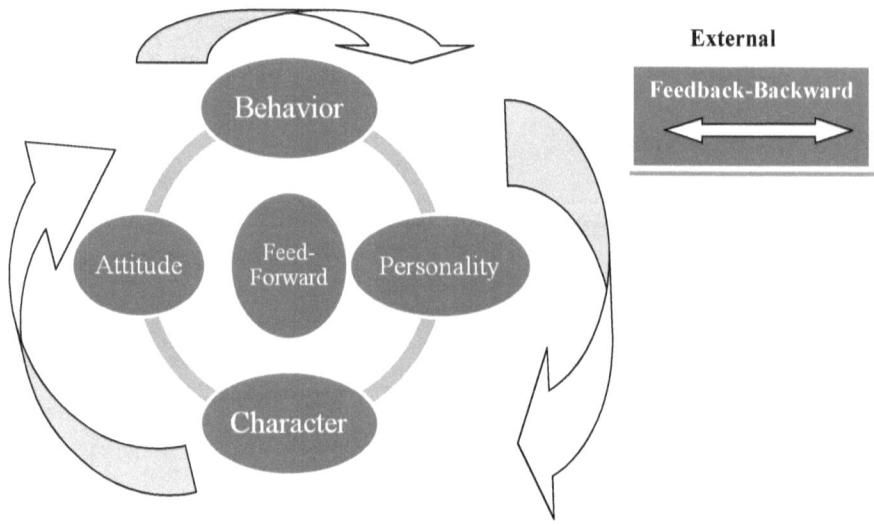

Behavior is shaped by External Feedback

2. The goal to live has to be directed by the aim to learn, which leads to the development of a behavior pattern in response to feedback from other people and the environment.

Human Systems Model 63

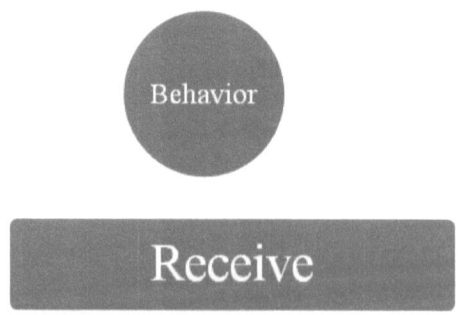

Behavior is anchored by internal feed-forward

2. A good behavior has to be set up to anchor how the child will grow in pursuit of three internal states: (1) personality, (2) character, and (3) attitude.

Human Systems Model 64

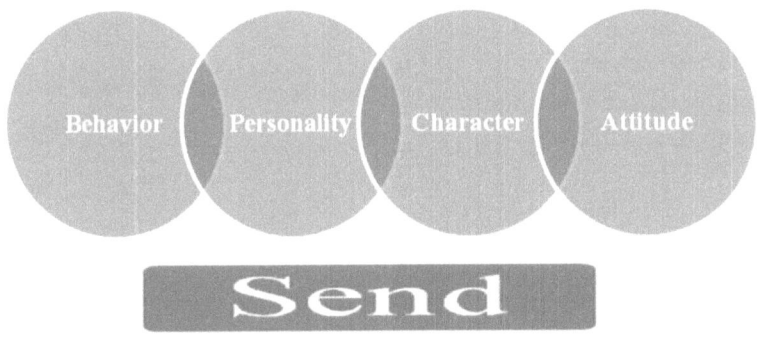

Parents search for feedback

3. The home has to be a safe place to build a strong behavior as a fixed state of responding to stress (parenting), which gives structure to the child's sense of the person they are on the inside.

Thank you!

This PERCS workshop began to take form on February 17, 2007. We tested the basic process cycles in our work with the California Community Colleges Foster and Kinship Care Education Program of Merced College: Building up a Child's Self-Worth Workshop.

Building Human Asset Process

When you work for and with a child in crisis, you need a system to connect a poor behavior to a process cycle, you need a process that sets up a reality-based time frame, and you need a strategy to help you learn through feelings of thought how to experience these ups and downs. Hence, you have to live each day to learn how to become a more informed asset for self, other people, and the child in crisis.

Closing Reflections

> How do you feel your brain?
> What do you think through?
> Why are you here?

Welcome to Our May 22, 2010 PERCS Workshop
Human Learning Goals Influence the Nature of Living, Learning, Thinking, and Responding as a Human System

Learning about the human system helps a child, youth, and young adult to live with a higher sense of self, as they grow to be more aware of their capacity and potential to act and to perform. Hence, he or she will act to move through contact, perform to move through interaction, and behave to cooperate as signs of care. This means he or she is learning to use brain, body, and sense events to participate as acts to live, learn, think, and respond to these feelings.

Working with Parents and Children 0-12

1. The focus for this stage: You will learn how to build up a child's personality and attitude.
2. You will learn how the brain has to be set up to send and receive information to learn from self, other people, and the environment.
3. You will learn how to set up a human learning goals program in the home to improve how a child feels and learns in school.

Progressive Investing Model 20

        Parent   Teacher

                     Child

Personality Development

1. The parent, child, and teacher (helper) form a vital relationship in the development of the child's personality (sense of feel for self).
2. Hence, the child's brain, body, and senses need to be set up in the home for the experience of the teacher inside the school.
3. The child will need a personal and a social sense of feel for self to work through an academic sense of feel for self.

Event 1
Human Activity
1. The experience of life inside the home is an event.
2. Hence, an event is a human activity.
3. In this sense, the purpose of the event to experience life inside the home is to prepare for life outside the home.

Goal 1
Human Learning Goals
1. The event to live through the experience of life inside the home is a human learning goal.
2. Hence, human learning goals are set up as the purpose in the act to learn how to live.
3. In this sense, the aim of a human learning goal is to learn from the experience of action.

Affect 1
Human Experience
1. The act to live inside the home to learn from the experience has an affect on how a child comes to think and reflect on life and learning.
2. Hence, the affect of human experience can produce good and bad feelings in the child's act to live and to learn at home or at school.
3. In this sense, the child has to be set up for the affect of human experience inside the home.

Family-Centered Learning
1. The child has to learn how to live inside a family as a human activity system.
2. The parent has to learn how to set the child up for the experience of contact outside the family system.
3. The child and parent share this need to learn how to live and learn together through family-centered learning, *goals*.

Progressive Investing Model 21

Event 1: Goal 1:

Affect 1:

Family Affairs

Attitude
1. Setting up the physical and mental aspects of the child's brain to receive contact and to process interaction is to develop their sense of feel for other people and environments.
2. The child, parent, and teacher are all physical, mental, and emotional.
3. In this sense, the child's attitude is shaped by the parent and teacher's state of mind.

Progressive Investing Model 22

Physical Mental

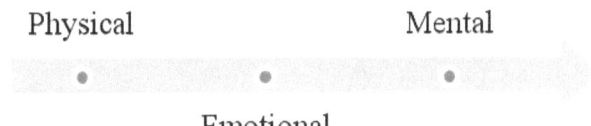

Emotional

Attitude Development
1. The parent is physical, which sets up the child's sense of feel for contact.
2. The teacher is mental, which sets up the child's sense of feel for interaction.
3. In this sense, the child's attitude develops from the ability to process parent and teacher interaction as models to learn how to cooperate with them as social and academic (psychic) forms of stress and pressure.

Progressive Investing Model 23

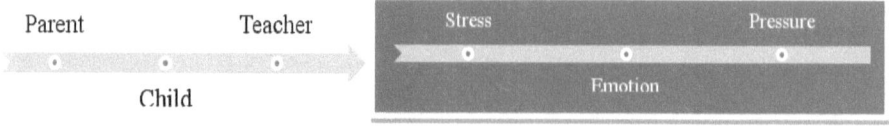

Event 2
Human Tension
1. The exposure of the child to stress through contact and pressure through interaction sets up a tension system *in* and *outside* the child.
2. Hence, a tension system is built up from stress and pressure at home and at school inside the child.
3. In this sense, the aim to expose the child to other people is to set up a sense of feel for human tension, stress, and pressure.

Progressive Investing Model 24

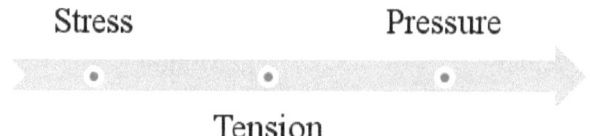

Goal 2
Human Thought
1. The experience of thought with real people in mind is a learned goal.
2. Hence, the exposure of the child's brain to other people is to set up goals to direct the flow of their contact and to help the child learn from the experience of interaction.
3. In this sense, the purpose of the goal is to lead the child *into* acts to think.

Affect 2: Act to Respond
Human Action
1. The act to think shapes responses to life inside the home, which have an affect on the experience of learning how to respond to contact outside the home.

2. Hence, contact outside the home affects a child's sense of feel for how to learn from the experience of interaction inside and outside the home.
3. In this sense, the child's act to think is an expression of how they interact in response to contact.

Thank You!

Our research-based practices serve as proof of the success we achieved through continued tests as the PERCS workshops took form. On August 13, 2007, we tested the system, process, and strategy, in our work with the Greater Sacramento Urban League. Data was collected using facilitated learning, activity-based learning, and responses to survey tools designed to identify staff needs in relation to their client systems to assess levels of crisis.

Helping other people learn how to help is a science. To learn how to help a sense of complex human systems is required to discuss mild, moderate, severe ranges of intensity. In other words, when you work with children in crisis, you need the ability to assess the child in reference to levels of mental retardation that can reduce their capacity to grow from where they are, mature from where they are, and develop from where they are mentally and physically and from the age and stage they are moving through. In measuring these optimal states, you need to know how to look for a child's highest levels of functioning.

In this sense, the human system grows more complex without a system in place to relate the regressive state to the progressive state, which then produces the data needed to learn from signs of a stable state through age appropriate activities. Helping then turns out to be a transaction between the helper and the client, his or her capacity and potential to become more informed as a competent service provider.

Closing Reflections

What is a reflection?
Who is a sign of stress or pressure to a child?
How do you feel things?

Welcome to Our May 22, 2010 PERCS Workshop

The word *delinquency* is a core focus to us, because it means different things to different families. To us, it means that a male is going through a difficult phase in the family system.

Hence, the problem of delinquency is as complex as the family system from which it spawns. We think it is much easier, then, to talk through the reasons for family decline that help to list ways to solve the problem of delinquency—running away from, cutting school, street fighting, selling street drugs, and so on.

Delinquency can be defined in terms of acts committed by a juvenile under the age of maturity, serious enough to require methods of control and treatment in a criminal context. Hence, to recognize a youth as a delinquent is to recognize a youth in crisis and a family that is in decline. Failure to obey orders in the home, failure to obey instructions in the school, failure to obey directions in the neighborhood, and failure to obey commands in the workplace is a refusal to receive contact, process interaction, respond cooperatively, and choose to participate.

Working with Males in Crisis 10-25

1. The focus for this stage: You will learn how to express feelings, sense experience, and the experience of experience.
2. You will learn how the brain has to be set up for contact in the home to learn how to interact with females.
3. You will learn how to communicate in feed-forward and feed-backward modes of learning.

Setting up the Brain to Express Feelings: 10-15

1. Learning how to read, write, compute, and speak sets the brain up to express feelings (Chomsky 2002) as thoughts.
2. In this sense, if you do not know how to receive information through the senses, then you have mixed feelings about the importance of school.

3. Hence, depending on your state of mind, you may be in crisis. This relates to stress at home, pressure at school, in response to a behavior issue, learning problem, or conduct concern.

Event 1: Language
How to Help
1. The use of talk in the home has to be set up to help males learn how to enter contact with females as social systems.
2. In this sense, to interact means the male has to be allowed to use the social skills being learned at home and at school in public.
3. This points to the male child's need for safe places to learn how to express a sense of feel for self as an actor from the inside out.

Goal 1: Feel
Help the Helper
1. The male child has to learn how to feel noises and sounds, signs and symbols as an actor to set up clear paths between the brain, body, and senses.
2. In this sense, the brain has to be set up to perform through the use of play to sense, feel, and focus how the body receives contact for social interaction.
3. This points to the process cycle of the senses to feel the noises and sounds, signs and symbols to focus how they respond to interact from the inside out.

Affect 1: Send and Receive Paths
Help the Helper Help
1. The male child has to learn how to send feelings forward as an internal feed path and receive thoughts backward as an external feed path: where dreams, visions, and awareness make up expressions of personality, what he feels; attitude, how he feels; and character, why he feels.
2. This means that the senses are aimed to feel for social interaction to focus acts to think through contact.

The Brain has to be Set Up: To Help
1. Males at this stage (16-18) have to know how to send, feel, and release feelings forward and receive, process, and respond backward.

2. This means as the brain sends feelings forward, the brain sets up paths to receive thoughts backward in search of feedback.

Progressive Investing Model 25

Feed-Forward   Feed-Backward

Send & Receive Paths

Setting up the Brain and Body for Sense Experience
1. Learning how to sense, feel, and focus the experience of sense data sets up the male's state of mind and awareness (Capra 2002).
2. In this sense, if you do not know how to process information through the senses, then you have mixed thoughts about the importance of school.
3. Hence, depending on your sense of awareness, you may be in a state of denial about the things you feel in relation to life at home and learning at school. You may withdraw, disconnect, and/or isolate.

Event 2: Cognition
Learning How to Live Together
1. The use of the brain to structure life inside the home has to be set up to help males learn how to cooperate with females in the process to live through thoughts of how it feels to interact with their contact.
2. In this sense, to cooperate means that the male youth has to be taught how to use the body to learn with females at home and at school in the neighborhood.
3. This points to the male youth's need for safe places to learn how to practice receiving thought backward in the sense of an actor in a search for understanding to learn how to perform.

Goal 2: Process
Learning how to Practice
1. Learning how to practice feeling for a sense of self, other people, and the environment is a search to learn how to process the experience in the real world.

2. This means that the experience is focused to aim the brain to live through, to learn through, and to think through the process of feedback in public places.
3. This points to the neighborhood where males need to have safe places to practice sending feelings forward to process their thoughts.

Affect 2: Receive and Send Paths
Learning from Experience
1. The learning process influences the way a person receives feedback and the way the brain sends feed-forward.
2. In this way, the brain, body, and senses express the meaning of information processing in the form of personality, who to feel for experience; attitude, when to feel for experience; and character, where to feel for the affect in experience.
3. This points to why the home, school, and neighborhood need to be safe places for males to practice receiving thoughts from the activity of sending feelings forward to learn from these experiences.

Setting up the Brain's Experience of Experience 19-25
1. As the brain sends feelings forward, the experience is felt, and when the senses receive thought backward, the experience is thinking through the experience.
2. In this sense, the brain experiences the experience (Capra 2000) of becoming physical and mental (Shoemaker 1999).
3. The body as a social and emotional system is used by the brain and senses to control and manage contact and interaction as the experience of experience expands signs of discipline and self control.

Progressive Investing Model 26

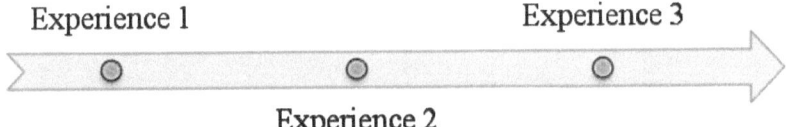

Event 3: Becoming Physical
The Use of the Body as a Physical State
1. The body is used as a physical state, because it releases a behavior pattern that describes how the brain sends feed forward and how the senses receive feed backward to live in the home.
2. In this sense, the body is used to feel the home for thought as a mental and social (emotional) state or condition of mind.
3. This points to the importance of reading the behavior pattern to learn how the body is being used in this physical state of becoming more informed or more ignored.

Goal 3: Feel the Release
You can *feel* and *think* about what you are in the state of what you are becoming
1. The physical body of self and the physical body of others is used to learn how to feel the release of information and meaning from reading contact and interaction to stages of cooperating.
2. In this sense, the brain sets the body up to release feelings to learn how to participate through the senses with contact and interaction to feel for signs of cooperation.
3. This point to the importance of brain recognition and the use of talk to feel the release.

Affect 3: Send and Receive Paths
You can *feel* and *think* about how the brain releases mental activity
1. The send feed-forward path is experienced in the contact between self and other people, and the receive feedback path is experienced through the interaction of self and other people.
2. In this sense, how you feel is experienced in the paths to live through, learn through, and think through the release of mental activity.
3. This pinpoints the importance of the mental state and the condition of mind, which produces the cognitive state.

Thank You!

Our research-based practices are also evidence-based. On September 10, 2007, we tested our information, knowledge, experiences, and reflections at the California Education Authority. We successfully tested our approach with curriculum and staff developers at the California Department of Corrections and Rehabilitation. Data was collected using facilitated learning, activity-based learning, and responses to survey tools designed to identify staff needs in relation to their client systems to assess levels of need. For instance, we asked the following:

1. What is the learning and support process for wards?
2. What is the instructional training process for teachers?
3. What is the therapeutic treatment process in relation to education or educating wards?
4. What is the communication system for connecting information, knowledge, experience, and reflection to the wards?

We explained in our request for a BHAM to improve learning and support services to poor, minority, high-need and special-need males to Superintendent Glenda Pressley. On
July 23, 2007, we discussed how, why, and who these males were that make up 95% of the Department of Corrections and Rehabilitations ward populations and the impact this problem has on the home, school, neighborhood, and workplace networks of children and families. We asked Ms. Pressley to learn with Save Our Youth

1. because the number of children with a parent in federal or state correctional facilities has increased by more than 100% since 1991,
2. because real and measurable progress has not been made at any point in standards-based education for males, and
3. because the cost to house one ward has reached more than $200,000 (CDCR 2007).

Ms. Pressley contacted Save Our Youth to meet with designated members of her cabinet. We are here to help improve learning and support services to poor, minority, high-need and special-need children, youths, and young adults by preparing better systems, processes, and strategies to inform and instruct the people in their family, education, government, and business systems. Our learning goals are shaped by the way we seek to inform and instruct parents, teachers, leaders, and employers who use the home, school, neighborhood, and workplace network as client, customer, or consumer-based systems.

Closing Reflections

> What does a male in crisis feel or look like?
> How do you learn to help?
> Why are our males at risk?

Welcome to Our May 22, 2010 PERCS Workshop

Human Learning Goals Influence the Nature of Living, Learning, Thinking, and Responding as a Human System

Working with Parents and Youths/Young Adults (13-18)

1. The focus for this stage: You will learn how to build youths and young adult's character
2. You will learn how to set a youth or young adult's brain up to experience cooperation.
3. You will learn to participate with a sense of feel for the environment in the home to improve how a males and females enter puberty, moves through young adulthood with discipline and self-control, and the experience of contact.

Character
1. Setting up the experience of contact for a youth and young adult is to learn the behavior, personality, and attitude patterns in their response, which shapes their character.
2. In other words, character is shaped by social and environmental experiences of contact.
3. For example, the act to make choices are based on how the experience of contact feels as observations of self and other people become known; they are social elements in the environment.

Character Development
1. The home is a social place, where people experience contact to learn how to interact with the experience.
2. The experience of contact activates the senses to react to the environment of home through feelings that result in thoughts about the people and certain decisions are made.
3. These patterns of responding to the people in a home are used to learn how to organize the way he or she observe, listen, learn, help, and lead the act to choose and decide how they will participate.

Event 3: Experiment
Signs of character: The ability to focus the brain under stress or pressure
1. The way a youth handles stress or pressure through a sense of self and other people is as distinct environments.
2. You experiment to learn if there are clear signs of a cooperative state of mind that explores the act to come to think, in and outside the home.
3. In this sense, these are signs that they are cooperating with the senses ability to focus how the experience of contact feels.

Thank You!

In compiling this data to improve our understanding of learning and support services, we met with the chancellor of the Los Rios Community College District on January 19, 2007, to discuss learning systems. We met with the president of Sacramento City College, Arthur Tyler, on February 19, 2007; and on September 16, 2008, we met with the vice president and a team of four deans on September 16, 2008. Our community-based learning curriculum improves the delivery of learning and support services to poor, minority, high-need and special-need youths and young adults because it informs the student how to learn by using the human system to receive, process, and respond to the experience.

Help the helper help systems are set up for human learning professionals involved in the health and human services field to make a difference through a congruent focus on ways to cause our home, school, neighborhood, and workplace networks to feel like safer places to live, learn, think, and respond in as we move through them (February 5, 2009, BHAM with the vice president of the Los Rios Community College District, Bill Karns). We prepare researchers, learning team members, consultants, trainers, and facilitators to help improve America's schools as part of the Building Human Asset Process.

Closing Reflections

What is a high-need youth?
How do you learn to be a good helper?
Why are you here?

Welcome to Our May 22, 2010 PERCS Workshop

Human Learning Goals Influence the Nature of Living, Learning, Thinking, and Responding as a Human System

Working with Parent and Young Adults (19-25)

1. The focus for this stage: You will learn how to talk to a young adult's brain.
2. You will learn how to use the experience of living, learning, thinking, and responding to home, school, and neighborhood events as the structures to motivate and transform these goals.

Cognitive Experiences Take Form

1. Goal-directed learning is a formal human systems mode for building up the child's sense of self, other people, and the environment (Slaton 2009).
2. Learning how to talk to the brain sets the human system up for the experience of contact and interaction.
3. Learning how to motivate a young adult to experience adulthood requires changes in the way you live, learn, think, and respond to them from where they are.

Talk to the Brain about the Roles of a Home, School, Neighborhood, and Workplace Network in Society

1. The goal to live inside the home is directed by the aim to learn how to leave, which leads to a behavior.
2. The behavior is measured by the way a person responds to talk about how to prepare their state of mind to leave home as a functional adult.
3. At this point, the structures of home, school, neighborhood, and workplace become key words that relate the discussion to these real life experiences.1

The Brain Has Formed a Sense of Knowledge and Awareness about Human Activity

1. A good behavior is needed to anchor how a young adult will display three states: (1) personality, (2) attitude, and (3) character.
2. Personality is related to life inside the home, attitude is related to learning inside the school, and character is related to thinking inside the neighborhood, because these are the places they have come to know.
3. The behavior they display represents their readiness for the workplace based on these three patterns of responding to contact and interaction.

A Pattern of Behavior is a Fixed State: Talk to the Person about the Feelings in their Character display

1. The home has to be a safe place to study a strong behavior as a fixed state to respond to stress, which gives structure to the young adult's sense of the person they are on the inside, the attitude they display from the inner and outer mode of feeling for a whole sense of self, and the character they are on the outside.
2. Talking to the brain, not the body: These human learning goals have to be set up in the home to focus how the body is aimed for other people to experience.

Talking to the Brain: Brings Back Memories of Changes and Consequences

1. The senses have to be used to bring the young adult to realize the condition of their brain in relation to the acquisition of status, resources, and wealth through the use of the body.
2. This is how you set a young adult's brain up to pursue college, the workplace, or a role in the armed services—by leading him or her to think; by bringing him or her to think about the purpose, meaning, and value of home, school, and neighborhood experiences.

Thank you!

We were contacted by Superintendent Carlos Garcia of the San Francisco Unified School District in response to our statewide cause-related marketing campaign and to set up a BHAM in late 2008. We met with Associate Superintendent Trish Bascum on March 16, 2009. Data collection is a continuous improvement process that allows us to advocate for parents and children in crisis at high levels of the federal, state, county, and local systems. On July 9, 2009, we met with Jamilla Moore, president of the Los Angeles City College, and received a tour of their learning and support services department.

We discussed the human systems information age. This new science-based approach unfolds as you make contact with our curriculum to sense, feel, and focus the way you choose to comprehend the need to change, adjust, adapt, or think through feelings. On October 29, 2009, we met with three officials of the Solano Unified School District. Educators have been trying to understand the home, school, neighborhood, and workplace network as parents, teachers, neighbors, and as workers so that the living, learning, thinking, and responding processes of human contact can be made more effective.

Many of these educators have interacted with children and parents with little knowledge on how the home, school, neighborhood, and workplace network makes up a child's learning system. Training on how to use the home as a living place, the school as a learning place, the neighborhood as a thinking place, and the workplace as a place to respond is a human systems research perspective being introduced through PERCS workshops to improve learning and support services to children, youths, and young adults that have been hurt by major life events.

Chapter 6

Reading to Your Brain: A New Human Systems Science To Solve Human Problems

The brain is wired to sense and receive data to improve how you read and interpret words and their meanings. When you use talk, the eyes and ears create more possibilities for the brain to experience ways to move through problems, issues, and concerns. Words create ideas on ways to explain how you feel and think through acts to read. Words help to redefine how you think you feel as they move through cycles of learning as tension, stress, and pressure adds to a deeper search for understanding. Simple words, root words, and words that connect to larger terms create a base from which to feel and think through your responses to writing.

These words are meant to be played with, tested, experienced, and researched to guide your journey to a healing perspective. Using each word is a practical strategy to transform a troubled state to a calmer and more balanced perspective on where to go from where you are as a human systems thinker (Slaton 2009), a systems thinker (Checkland 1999). Do you think before you feel, or do you feel before you think? How do words move through you? Do you always resist, or do you just deny yourself their experience? You will get to tests your capacity and potential to sense and receive information to act on it, live through it, learn from it, think through it, and respond from it.

Learning how to sense and receive words moves you beyond tests and beyond time. To feel life from where you are allows you to explore how the next step ought to feel as the brain begins to take the process over. You have to trust the person you are, to learn more about the person you want to be. Then when can feel it—save it to memory and reflect on how the experience feels, and you are thinking through acts to live, learn, think, and respond to feelings from where you are.

Healing Time

Before you begin the act and process to read how the human system thinks with the brain in mind, let us tell readers why this is healing time. The reader will be asked to consciously use time to structure how the brain functions to use the body to sense, feel, and focus characteristics of living, learning, thinking, and responding to tension, stress, or pressure. Healing time is the act or process to improve how you practice and participate to learn how to use the brains body to heal.

We use talk through the use of special words to discuss state of mind, systems of the brain's body, physical, mental, emotional, and social learning as effectively as possible. We have used Human Systems Research to assess the human functions of people that have been hurt by major life events in the home, school, neighborhood, and workplace networks to learn from the people that have been most effected by human problems.

We have studied family decline, school failure, deviance, and unemployment across the nation going back to the drug epidemic that began in the 1970s, on behalf of children, parents, and grandparents. We have observed, learned, listened, helped, and led therapeutic projects in efforts to improve the lives of children that have been abandoned, abused, neglected, and subjected to drugs while in utero by parents who abused drugs and may continue to abuse drugs as they age.

This is why, in our comprehensive systems of family, education, government, and business affairs, people that are hurt need to learn how to help the helper help him or her screen, assess, and measure how well they can enter contact, interact, cooperate, and participate in their home, school, neighborhood, and workplace networks that will decide their successes and failures.

To solve human problems of this nature, we offer fifteen sessions of talk, using words that ought to flow through the brain's body to explain the human systems approach to help heal the person over time. We talk to the brain to help the person think about how to use the brain, and how the brain has to be built up to learn to live with the self, other people, and the environment as carefully as possible.

This work is important: People's families are at risk, people's educations are at risk, people's government services are at risk, and people's businesses are at risk when there are so many people being upset by human problems. In other words, a great number of people are not able to move through these systems without anger, fear, or anxiety; frustration, disappointment, or hostility without being violent, aggressive, and intimidating toward self, other people, and the environment.

Most of which, at some point, give up, quit, resist our help, and go from mild, to moderate, to being severely disturbed. This is a Progressive Investing perspective that has been designed, from more than fifteen years of work with children and adults in their home, school, neighborhood, and workplace networks. We study how people that have been hurt by major life events learn. When a person is able to process words through the use of talk and the practice of participating, they are helping to heal themselves through the act and process to improve their performance.

Session I

Serve time. We begin with you to learn with you, how to help you serve you. We want you to know what time is, to learn the value of your time. The time you have to live, for instance, has to be set up for you to serve. When you are set up to serve, the way you release your services create an event.

If you are the event, then you have a goal to serve in mind. When you take the time to set up your role, you create time to satisfy your need to serve. You think about the time you have to serve. You think about the way you want to serve over time. To form a service to self and others, you have to think through your goals.

The goals you set up will have an affect on how you form an attitude. When you choose to serve time with a goal in mind, the experience of the service and the satisfaction you feel or fail to feel forms an attitude. An attitude is a point of view that concerns how you feel in life at a certain point and time. How you live and the way you think is part of this attitude. To serve time, your body and mind needs to be set up to live with you.

When you were born, you became an event. In other words, you became an experience to self and the other people who could feel the life run through your body. As such, you were born with a goal, and you have a goal in life to live. If you have not yet done so, then now is the time to set your body up to live. Your body serves as the physical make up of who you are as an event.

Your first goal in life is to seek the need to be a service to self. By this, we mean, live each day to learn how to use your body to serve your mind. If your mind is not connected to the way you choose to use your body, then you may not be satisfied with the way you live. The body and mind make up a service to self. In service to self, you must live each day to learn how to set up the kind of goals that will lead you to a unified sense of body and mind. This is a good way to live.

Hence, you change a bad view of life into a good way to serve time. If you choose to seek ways to get your body and mind in order, then you will. What we do is help you get started. First, the body and mind is easier to set up to live and grow when you seek to use it as a human system. A human system is what your body and mind looks like when you are able to sense, feel, and focus the way you live each day.

When we say sense, we mean to say your body is used to feel things through the eyes, ears, nose, skin, mouth, and atmosphere. In contrast to the body, the mind is aroused by levels of focus that are acquired through the things that come in contact with the body. We call this a process of body, mind, and environmental contact that is transferred to the brain through nerves.

Nerves run throughout the whole body as a nervous system. The nervous system sends and receives messages through the use of the brain as the central nervous system of the body. The word *mind* is synonymous with the word *brain*. The words *body* and *mind* are synonymous with the words *human system* in this same regard.

When you think about a human system, you think about the body and mind as a whole person. This is an important point to note, because your body cannot live without you, and your mind cannot live without you. Your body has to learn to serve you, your brain has to learn to serve you, and you have to learn how to serve time in your human system. If you lack this sense of self, then there may be disconnect between how you feel things through the body and use the body to transfer those same things to the brain.

Take for example, a bad attitude, which is a bad feeling related to some experience that might cause you to resist being helped by us. This means that your brain is not open to the experience of us, as helpers, and that the way you feel on the inside affects the way you feel about us on the outside of your body. This is what we call disconnect. You may not be allowing your senses to transmit data to your brain, and your brain may not be allowing your feelings to be focused. Feelings are the good and bad states of mind that may result from contact.

Through good or bad feelings, you have to focus how the experience feels to release the act to sense and feel the contact of the body, the mind, and the environment. The body learns how to use the brain to live with a higher sense of self and the atmosphere of nature. We connect the event of life to the service of self; the goal of life to the attitude that forms a sense of self; and the affect, the event, and goal has on his or her human system's need for a sense of satisfaction with the state of their body and mind.

Progressive Investing Model 27

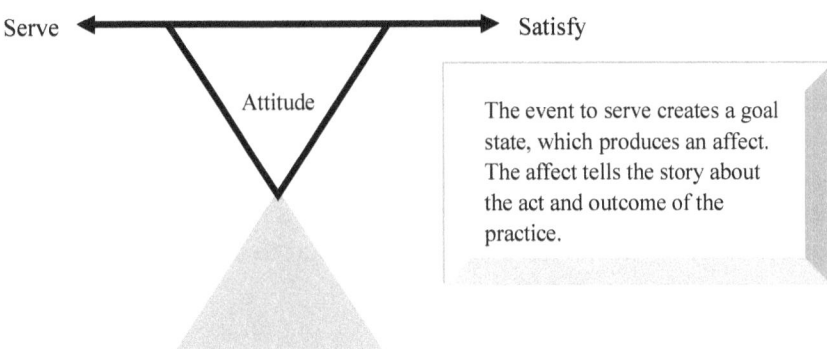

Session II

Care costs. We look at you to learn from you what is the state of your body and your mind as a system. A human system is like a network, where nerves run throughout your body: connect to your head, heart, lungs, kidneys, abdomen, and muscles; brain stem, spinal cord, neck, shoulder, chest, stomach, buttocks, thighs, legs, knees, ankles, feet and toes. You need to know this, because you need to care for the way these connections help you to live, act, and learn.

This is why we say care has a cost. You do not just live; you have to be able to learn how to act to live. To take action to live, you have to be able to learn how to live with self, other people, and the environment. In other words, you have an inborn need to learn how to act, to learn how to interact, to learn how to cooperate, to learn how to live.

Care has a cost. We associate these cost with the time it takes you to come to realize the needs of your body and mind to work for you as a human system. It takes time to move from a state of living to a state of learning how to serve both your body and your mind as a human system. This is why we use the word *care* to tell you there will be costs to learn how to use your body and your mind from where you are as a person.

First, like a service to self, care is an event that has to be experienced by you but in relation to other people. Signs of care need to be released from you, in the sense of how you use talk, to make contact, and to interact with other people, to learn from the inside of your body, to live with the world outside of your body. When we say released, we mean the brain's response to contact needs to be free to communicate. Not resist, but move in the direction of care for self, who has to learn, and care for other people whom have to learn also, how to live with you.

It cost to live without a sense of self, and it costs to live without a sense of the people. Hence, care cost. You set up these costs when you do not live with care. You may not want to think, or you may stop trying to do what it takes to satisfy your needs. You have a need to live with a feeling of focus, which gives you the strength to compete. When you step outside the self to free a response through the use of talk, you can feel the other person's energy stare back at you.

You can feel the energy, or you can resist the energy, but your body and mind is sensing, feeling, and focusing a response as you use talk. If you resist the need to feel the energy, then your use of talk will not seek to penetrate the surface of the other persons, body, and if so, signs of care must be plain enough to share. This means, you have to be open to their contact, and they have to be open to your contact as well. This is why you fill the words you use to talk with, with signs of care to reach his or her brain, hence, obtain the reaction you need to learn how to live with other people.

The goal of an exchange is to show clear signs of care for the contact, even if you feel bad, to learn how to move through it. Hence, the goal in an event to learn is to focus the way it feels to have feelings run throughout the whole body. Your human system is set up to use talk to deal with tension, stress, or pressure. This is why there are times when you feel the words just did not come out the way you think they should have. When you use care as an event to focus, then the affect of your use of talk to release energy has a clear goal.

You want your words to be heard; they first must be received. If your words are filled with bad feelings, then the other person will have to set up a control and management system, or a resistance to your contact, because it seeks to arouse bad feelings. This is why we ask you to learn from where you are, and not from where you think you ought to be, in the problem of living and learning how to use care to focus the way talk comes out of your mouth.

We could use your eyes, ears, hands, or feet in the same sense of how we have used your mouth and talk, as objects to learn through the use of care. Instead, we help you to make this an easier thing to grasp through the words *human system*; meaning, they are equal in the sense of devises used to communicate events, goals, and affects. This in turn makes it easier for us to talk about the costs of not caring to sense, feel, and focus to live and to learn how to use your body and mind to communicate as a system.

Only humans can communicate using the self, other people, man-made products, and the environment to send and to receive messages as a systems response to tension, stress, or pressure. There is no better time than right now to learn how to care, for who you are. You are a human system made up of a body and mind that lives to learn how to live with self, other people, and the environment. If you are here to help, then you care.

When you use talk from a point of care, your words focus on what it means to be in a body and work with a mind that thinks to live and to learn. Hence, we say the body lives to learn, and the mind learns to live. These are two vital points in the event to care and the goal to use talk, which has an affect on how you focus. If you are here to help, for example, then the sense, feel, and focus, cycle is open to living and learning as a human system that cares to feel the world.

This means that you are open to the experience of good and bad feelings to learn how to live, which in turn means that your brain is in the lead role of the act to think. Three things have taken place: you live in your body to learn how to feel the world, you learn how to live to allow your brain to take the lead, and you think to focus how you act through signs of care. You are here to serve time by grasping this sense of what it would costs, not to care for how your body is set up to work with your mind.

Progressive Investing Model 28

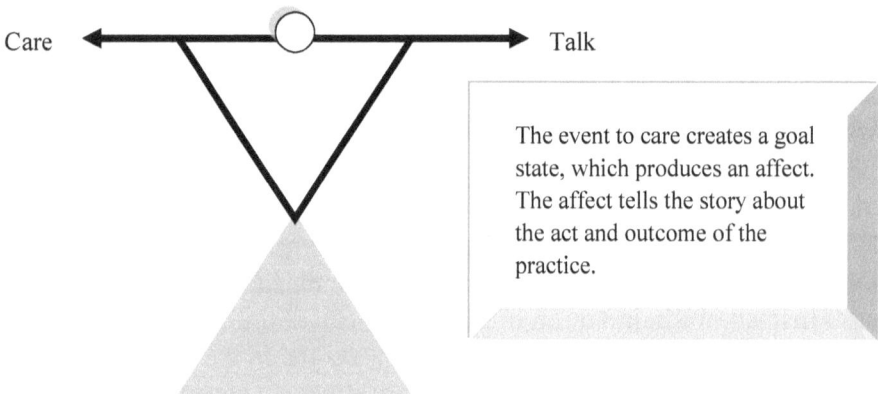

Session III

Feel the world to think. When you are here to help yourself, feel the world, you are here to think. This means that your senses are open to how it feels to live in your body, learn with your brain, and think through the environment of your human system. To feel the world makes up an act to live, learn, and think. These are all acts to live and learn to think. The human system is in the think mode, to act to live and to learn all the time. When you wake up for example, your eyes open to feel where you are and your ears hear to feel what your eyes cannot see.

Your human system seeks to feel the self, space, time, and any signs of energy. We say this is how the human system lives to learn how to think through the sense, feel, and focus cycle. When you lead with your brain, the brain senses, the brain feels, and the brain focuses your body in a cycle of behavior that is open to change based on certain results. Yet again, your human system only works like this when you are here to help. Hence, if you are here to help, then you are here to serve.

If you are here to serve, then you are here to learn how to use the body and the mind over time to compete with other people. Hence, if you can feel the person, you can think with that person in mind. This means you can set your human system up with the type of behavior that will lead to feelings of care: potential, control, strength, and influence. In other words, you are able to manage the flow of energy through your human system in response to any pressure. A behavior that does not flow from a state of care may be filled with feelings that lack a connection with self, other people, and the environment.

A behavior that does not flow from a state of care may be filled with feelings that lack a connection with self, other people, and the environment. This is why you act to feel the flow of energy, to manage the effort to show signs of care to learn in relation to the possible effect on you. You check your behavior for signs of tension, stress, or pressure in this way to learn through the feelings that are provoked. Hence, the act to feel is a focused effect that provokes signs of care.

For example, when bad feelings sway your body and mind, you are more likely to seek to intimidate, bully, and lash out aggressively at the people who try to interact with you. We say this happens when your body is loaded with anger, your mind is loaded with fear, and your environment is loaded with anxiety. Anytime you do not have answers or sufficient knowledge to respond to a situation may increase the risk that these behaviors will be released. They mean "move away, turn away, do not try to talk to me, and do not stare at me, I do not want to think right now." Your human system is filled with tension, stress, and pressure.

The power to be you, feels like tension. You stretch, you sigh, you look around, and you check to learn how your body feels. You hear a voice, you see a person, you shake a hand, and your brain wants to know more. This is stress. Stress feels like tension, except that tension lives inside your body as part of your message system. Stress on the other hand grows out of feelings that relate to interaction between you and the outside world.

For the most part, we say human relationships carry stress since they require the use of care, talk, and focus to interact socially and competitively. When we refer to the environment, we mean the natural world. Hence, stress would include man-made products and services. Pressure is what you feel in the atmosphere as wind, heat, rain, dust, and so on. In each of these situations, you have to feel the tension, feel the stress, feel the pressure to learn how to live with self, other people, and the environment.

We call this process an act to learn how to live and think through contact. You will learn how to grow with a behavior that is open to feeling the world to learn because you believe in the way your human system performs with thought through action. You can feel your brain in the lead, because how it feels to have a feeling system is under the control of your body and mind as they work in tandem to serve you. For example, as tension builds, your eyes perk up; as stress builds, your ears perk up; and as pressure builds, your mouth perks up in search of ways to release the energy you need to learn how to live through these exchanges of contact.

Your human system is organized to sense, feel, and focus tension, stress, and pressure to move feelings through a process of managed care that releases a behavior, as a pattern of how you think in complex situations. Your human system thinks all the time. You will feel the benefit in learning how to manage and control your behavior all the time. You will have feelings flow through your human system all the time. Any contact with the human system, spurs change, since your brain seeks signs of life from the inside out, all the time. You have to be ready to serve all the time.

Your ability to think controls how you feel the world to manage how you change from one cause to the next. When you live to learn how to learn with a body and mind that is bored of being divided, the act to live is connected to the act to learn; you can make sense of the tools inside your body. You body and brain does not have to be in unequal worlds. How does it feel to work with a body that has all these tools in it, to control your behavior, through feelings of care? You live to serve, you learn to care, and you feel the world to think.

We think what the world needs right now is more human systems thinkers. Children, youths, young adults, people who want to be able to deliver a service, people, who want to be able to show signs of care, people who want to be able to feel the world to think, and people who want be able to experience a higher degree of satisfaction, focus, and feelings. The human world is made up of real people. Hence, what matters most to us is how we prepare you to be a human systems thinker.

Progressive Investing Model 29

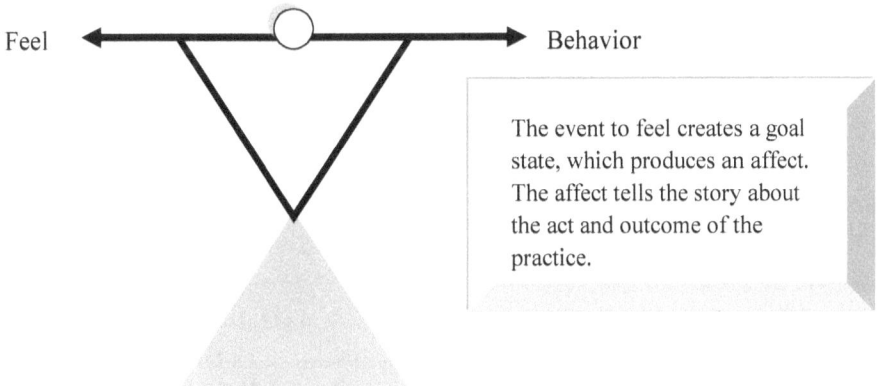

Session IV

Trust is a path to respond. We may not work for you if you lack trust. We have to believe or we need to feel that you believe we care and that you want to care. You will be asked to help build our confidence in you, as you are able to manage the flow of energy we rely on to help you. Trust is a path to respond.

If you have lived long enough, then how you live in a home, learn in a school, and think in a neighborhood you trust is a path to respond in the workplace. You learn from where you are to where you want to be, by learning how to use these paths. If you take the time to learn how to live in a home, learn in a school, and think in a neighborhood, then trust is in the path you choose to respond in the workplace.

If being able to lead a family is important to you, if being able to use an education is important to you, if being able to participate in government is important to you, and if being able to focus on business is important to you, then trust is a path to respond. We are asking you to trust us based on three variables we have already reported. Trust is based on how you sense, feel, and focus to respond to tension, stress, and pressure.

How much tension do you feel? How much is the stress to feel at home, in school, and in your neighborhood? How much pressure do you feel when you connect these feelings to the workplace as an ultimate objective? Can you see yourself meeting the need to succeed without a plan on how you will move through these places? Are you here to serve time? Are you ready, because to live in this world, you have to care, and care has a cost. Either you learn how to feel trust, or you will not have trust to work with us.

To feel the world to think, you want to be able trust your human system. You want to be able to move through your home to live with a plan that allows you to grow through good and bad feelings. You want to be able to move through a school under equal terms to learn how to deal with different people in a competitive environment. You want to be able to think through your neighborhood relationships to practice observing, learning, listening, helping, and leading yourself toward the ultimate goal.

Living in a home, learning in a school, and thinking in a neighborhood are three things you must master by your thirteenth birthday for sure. You have to be trustworthy. You have to be worthy of the trust of the people who live, learn, and think with you in your home, school, and neighborhood situations as you move through them. You have to have confidence in what you bring to the game of living, learning, and thinking with the people who compete with you or against you, in these places to respond and compete for roles in the workplace, which mean something.

Trust is a path to respond, but the path you take has to be more reliable as you move through these pathways to learn. You may not like the home you live in, you may not like the school you learn in, and you may not like the neighborhood you think in, so what is your plan? What is your plan to deal with the people who make you feel bad, without hurting yourself, other people, and the environment any further? How do you plan to live through poor experiences that cause you to lose confidence in your freedom to sense, feel, and focus?

You have to build trust, and you have to learn how to build a sense for finding people that are trustworthy or worthy of your trust. Hence, the tools you use to make this journey feel real and have to be reliable to you. By the time a person turns thirteen, life is coming at them fast, hard, and steady. He or she is going through physical, mental, emotional, and social changes. He or she is meeting more people from different backgrounds and experiences. Hence, he or she must learn from them to refocus their goals and aims. He or she is under a rousing degree of tension to find a sense of self, which feels more comfortable with these changes.

This is why we say it is the people that carry the problems of living, learning, and thinking alone and with other people from simple, to complex, to serious problems—getting along with other people in their home, school, and neighborhood situations with them. Hence, we say, you need a sense of the people you will interact with from one day to the next to assess them for their reliability and feelings of care for self and others.

If you get hurt, you want people around you that are not afraid to feel your pain. You want people around you that are not afraid to show you signs of care. You want people around you who live to serve as a group of people with a will to help each other live, learn, and think. In other words, you want to help build more human systems thinkers that are prepared to live with you and without you to meet these demands for people we can trust in roles, where being trustworthy means they are reliable.

You can feel trust. You can experience being trustworthy. You can practice being reliable. Trust is a path to respond you build up from your contact with the people who interact with you in your home, school, and neighborhood situations. You learn how to compete, where to compete, when to compete, why you need to compete, and what to compete with through these pathways.

To arouse these pathways, you have to feel them. You need to feel them to learn how to act through them to manage the person you may become as you move through them. You learn how to trust yourself through these acts to participate in these pathways. Learning how to live in a home, learning how learn in a school, and learning how to think in a neighborhood creates a higher sense of self in the exchange of contact where you can feel trust. You can feel a sense of being reliable. You can feel these feelings of being trustworthy.

Progressive Investing Model 30

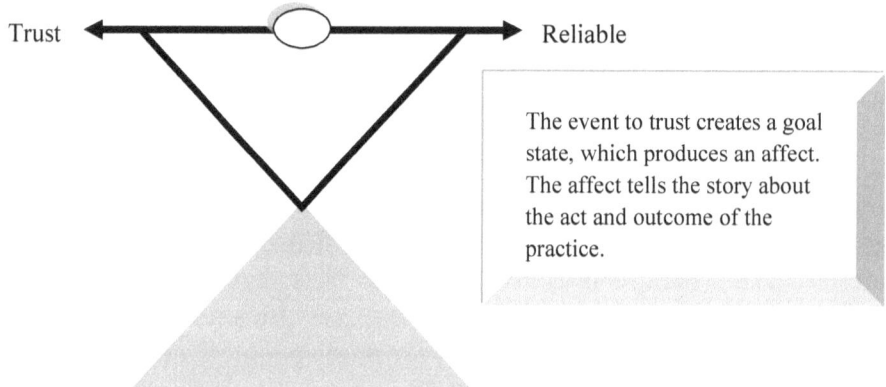

The event to trust creates a goal state, which produces an affect. The affect tells the story about the act and outcome of the practice.

Session V

Know who you are when you think. You have to be able to think to have a sense of who you are. When we ask who you are, you may not just say you are somebody. You have to be able to think about the things that you can do to participate in your home, school, and neighborhood networks.

To think you are somebody is not enough. You have to put in the work to become somebody other people think about doing things with to live, to learn, and to think. Hence, we say you need to know who you are when you think. To respond, you have to be able to perform in home, school, and neighborhood situations where your capacity to think makes you a more reliable person.

You have to be honest: What are you like? When you do not think, what are you like? You have to be able to trust your feelings. If you care, then honesty is the part of you that seeks the opinion of others through your acts to live, to learn, and to think as a human system. What is a human system that cannot think? To think, you have to have a sense of, a feel for, and a focus on who you are, what you may become, and how you need to act.

To move your need to act, we have asked you to think about why you are here, where you are at your best, and when is the best time to be you. From this perspective, we have moved through the act to serve time, the act to show signs of care, the act to feel the world, and now, the act to be honest with yourself about who you are when you think. If you are being honest with yourself, then you are being loyal to your sense of self.

You want to connect these acts to the experience of learning who you are as a thinker. Are you honest enough and are you loyal enough to probe who you are, in the sense of what you may become as a thinker? You have to validate how you feel as a thinker. You have to validate how you focus as a thinker. You do this work to connect how you sense, feel, and focus as a thinker.

This is why be honest, be loyal, and be able to validate your ability to know how you are when you think. In other words, are you able to receive contact between self, other people, and the environment? Are you able to process contact between self, other people, and the environment? Are you able to respond to cooperate with self, other people, and the environment? Now, are you able to do these complex acts in relation to how you choose to live in a home, learn in a school, and think in a neighborhood?

Without these skills, you are not able to think through these human relationships. For example, how do you deal with the parents, teachers, neighbors, and leaders you meet in these situations that may or may not feel good to you? Do you seek to ignore them, resist them, deny them or displace them? You have to be honest. What do you do to help or add hurt to the situation?

Every experience counts for something when you know you are thinking through the act to live, learn, and think to respond. If you know you can trust your ability to think, then you move through the home, school, and neighborhood learning system with a sense of confidence about who you are and a feel for what you may become with a focus on why you do this work to live, learn, think, and respond to contact, interaction, and cooperation.

Cooperation is the most vital ingredient, because you have to be open and flexible to live, learn, think, and respond as a human systems thinker. The ability to cooperate means you can participate in acts to live in a home, learn in a school, think in a neighborhood, and respond in a workplace as a whole practice. However, you have to be honest with yourself, because you have to do this work yourself.

If you have good parents, then that is a plus. If you have good teachers, then that is another plus. If you have good neighbors, then that is yet another plus. Still, if you do not have good parents, teachers, and neighbors, you have to use these skills to live through bad experiences to find the leaders in these pathways that are there to help you move through the home, school, and neighborhood networks more successfully.

Leaders are the people who do not have to help you, but they try to help you, because they may have been in your shoes from another point of view. Hence, if you are not ready to receive his or her contact, then you are not ready to process the words he or she will use to reach you. If you are not ready to respond to his or her cooperation, then you are not thinking to serve, care, feel, trust, or be honest enough to learn from him or her if they have these aims, these skills, these feelings, a sense of honor, or if they know what it feels like to be where you are.

You have to learn how to live through contact that may feel bad, but when you trust your human systems ability to channel these feelings for you, you increase your knowledge of the skills you need, to work in each complex situation. If the contact cannot be received, then you need to know why. If the interaction cannot be processed, then you need to know why. If the cooperation cannot be responded to, then you need to know why. You live in an experience-based body we call a thinking system.

Your body responds to experience through the act to live in a home, learn in a school, think in a neighborhood, and focus on the workplace, which leads to a systemic response to tension, stress, and pressure. This means that you take action to learn how to function in these places with intent to study the self, other people, and the environment. Hence, you have to be honest and loyal, because you have to validate your ability to function as a human systems thinker.

The validity of a human systems thinker can be easily assessed by how well you are able to sustain the positive levels of contact, interaction, and cooperation needed to learn how to compete for success and recognition. You compete for success through your ability to sustain acts to live, learn, and think; to gain the recognition you will need to experience in home, school, and neighborhood situations; to serve, care, feel, trust, and be honest about who you are as a human systems thinker.

Progressive Investing Model 31

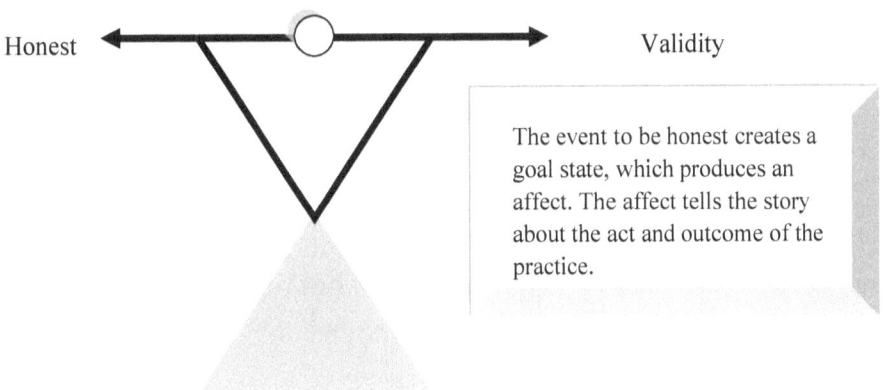

The event to be honest creates a goal state, which produces an affect. The affect tells the story about the act and outcome of the practice.

Session VI

Accept this chance to change the way you act. If you let yourself receive this opportunity transform the way you process your feelings to work with us, you will learn from the way we use words to talk to your brain. The brain is the part of your body that makes all these transactions doable. Accept this chance to change the way you act.

We will ask you to move from a poor behavior to one that works to help you serve your needs. We do so through a field of experience that tells us it is easy to say, but harder to do. This is because a poor behavior carries a pattern of negative contact, interaction, and levels of cooperation between the self, other people, and the environment. Hence, we use these words to set up a learning plan.

Are you here to accept this chance to change the way you act? You have to act through a bad behavior, because you need to work through how you feel, and other people need to work through negative thoughts about you. When you act to live, you live to change the way you act or behave. This is an opportunity to learn to use your brain to change how you feel on the inside, by acting as if you want to be good to the people on the outside. Your brainwork has to be in the front, not in the middle or in the back of your act to learn.

No one should be able to say you do not know how to act. An opportunity is a chance to learn how to accept the way you change, when you choose to act through your human relationships. We want you to have a chance to gain access to the best places to live, learn, think, and respond to contact with people. You must want access to the best homes, schools, neighborhoods to make this plan work for you. You have to want access to the best people to live with, learn with, think with, and to respond to. You want access to the best.

You have to show people how well you can act to gain access to these places that improve your ability to compete and feel the world. When your behavior is under your brain's control, you seem more open and flexible and easier to work with in a complex situation where there are formal rules of engagement, like in a home, school, neighborhood or workplace. When you gain access to these places, you have to show signs of care for the peoples who share them.

You need to decide, do you want to serve or act as if you have served? If you want to be accepted, then you have to choose to act as if you care or choose to care because you need to feel healthier. You have to gain access to your own brain to learn how to focus your act to serve and care about the access you can achieve. It is really up to you. To feel the trust of other people, you have to act as if your behavior is under control. You have to sustain this act throughout your whole life.

Otherwise, you will fail to experience the contact needed to learn how to function with a growing sense of self, other people, and the environment to compete. When we say compete, we mean with your sense of self-control, and then to live, learn, think and respond at the highest levels of self-actualization that are possible for you. Contact is the opportunity to experience the act to change and be changed by the reaction you receive from having acted with intent to learn from the exchange of contact.

Successful contact with self, other people, and the environment can gain you greater access to safe places to practice learning how to live, learn, and think through the act to accept this chance to change the way you behave. You have to know how you behave and admit that you need this talk to help you think these things through. You have to be honest with who you are. You have to want to lead with your brain and not with your body. You need to lead mentally and then physically.

The way you act and the way you behave are two complex variables. Each variable, however, is dependent upon your brain's powers to change or control the way you use the human system to communicate. In the act to live, learn, think, and respond, we expect you to change based on how you sense, feel, and focus through the experience of contact. To change the way you behave to live, learn, think, and respond, we expect you to use discipline through a sense of self-control, which tells us your brain is in use every step of the way.

Your behavior cannot be closed to contact; it has to be open to change. A behavior that is open to contact is in a state of change based on how the brain channels these feelings through receive, process, and respond cycles. For example, a mental behavior focuses how you feel, so your personality or sense of self is the core path you use to learn how to act at home. Character or your social self is the core path you will use to learn how act in your neighborhood. Attitude or your emotional self is the core path you will use to learn how act at school.

Your behavior will take in each of these core paths as equals to sustain your act to display acceptable changes in personality, character, and attitude. Each of these words will relate back to the structure of competitive contact, interaction, and cooperation in the goal to participate as the person you are. When you manage your behavior, it is a life skill; when you manage your personality, it is a personal skill; when you manage your character, it is a social skill; and when you manage your attitude, it is an academic skill.

If you accept this chance to change the way you act, then you will. With these tools, you will learn how to organize your sense of self to serve, care, feel, trust, be honest, and accept this opportunity to set up your behavior. A behavior that is open to contact is flexible, because it functions under your brain's control over how your body behaves to learn through the act to live. Hence, the person that you display on the outside looks, feels, and acts like a thinker. The qualities that you display on the outside serve as proof that you can grow, mature, and develop as a competitor.

Progressive Investing Model 32

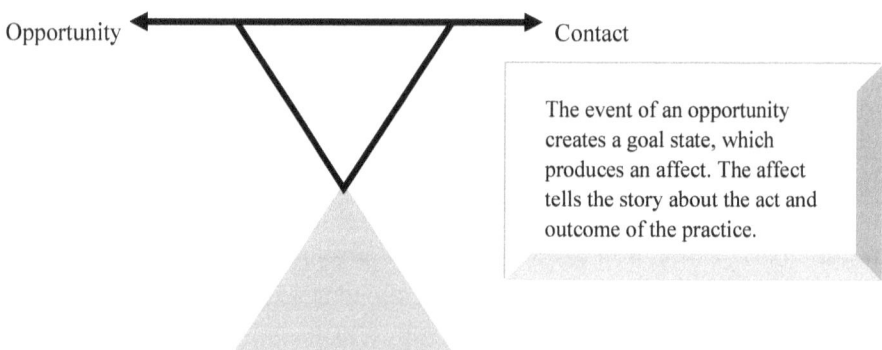

Session VII

Learn to use your brain to change your mind.
If you study how we set words up to talk to your brain, you can feel them. If you feel the words, you can grasp a sense of our need to be heard through your need to hear us. You have to be able to help us reach you. This is how we ask you to help us help you. Learn to use your brain to change your mind.

We have used the word *experience* throughout our talk with you. *Experience*, to us, is the most powerful word in the plan to help you learn to use your brain to change your mind. This is because, when you use your senses to feel the world, you are using your body to experience how the world feels. When you can sense, feel, and focus this intent to seek experience, the whole act is mental.

The experience is felt in the brain, with intent to channel how it feels to a path of focus to change your mind. The experience of sensing, feeling, and focusing contact can be good or bad, which depends on the mental state your mind is in at the time. We learn from experience how to use our brain's power to change how we sense, feel, and focus to talk to your brain. This is because, we know you can change your mind about how you use your body to sense, feel, and focus with us to learn from us.

Our goal is to observe you, learn from you, listen to you, help you, and lead you to learn from us how to use your brain to change how you receive, process, and respond. Change is an inevitable process that is going to take place regardless of whether we want to change, you want to change, or nobody wants to change. The experience of contact changes us all. Hence, our goal is to make you more mindful of the practice we use to manage and control how we change with time.

If you learn to use your brain to change your mind, you will do this through how you practice, the act of learning from experience. For example, as we experience you, we learn from you how to experience you. As we observe how you act, learn from the way you act, and listen to the way you act, to help you act, we have to practice following your act to lead. In each step, there is the experience of being an experience in relation to other experiences that will all relate back to our initial contact. Such as how you take in these words and pass them on to reach your brain.

If your brain is not in the lead, then your body will not relate to us, reflect on us, or represent signs of these experiences. When your body is in the lead, your brain has to wait until the transaction is over. For example, the experience of our contact may arouse and cause a reaction, but not brainwork. This is why it is vital that your senses be on to feel any and all contact to focus the act to become more mindful.

Your body has to be set up to transfer the mind to the point of contact to experience the act to observe, learn, listen, help, and lead. You have to learn how to practice using your brain and your body on equal paths to change your mind. To control your mind, the brain has to be on task all the time to learn from the experience of contact, how to experience the type of changes that will satisfy, focus feelings of being trustworthy, loyal, and accessible through this practice.

If you act to live in a home with people, if you act to learn in a school with people, if you act to think in a neighborhood with people, then these acts become a performance. Your performance becomes an experience, and your experiences of these experiences become the path to learn how to use your brain to change your mind. This way, the experience of self, other people, and the environment counts for you.

To sense, feel, and focus this way, your brain has to be in the lead so that the whole body is protected from hurt, hurting others, or destructive acts. When the brain is in the act to sense, feel, and focus a sense of self, other people, and the world, your body has more self-control. This is because with the brain in the lead, the body needs to learn how to observe things through the use of the senses, help to control things through the use of the brain, and listen to use the world to receive, process, and respond with change.

The experience of the brain being in the lead create changes in all of us. Most of which we live through with the goal to perform with a higher sense of self that gives us the strength to practice experiencing these experiences long enough to learn how to use them to serve, to care, to feel, to trust, to be honest for the opportunity, and to experience human system thinking.

How do you think with your brain? How do you learn with your mind? How do you experience change? You learn to use your brain to change the way your mind comes to think. In each step, you have to act with your brain to use a home, a school, and a neighborhood to focus how you will use the workplace to meet these goals. This plan may seem simple, but we have learned from experience that this is a complex learning plan. Each act calls for experiences with contact. Each act calls for you to grow, mature, and develop through these experiences.

Are you here to help us help you learn to use your brain to change your mind? It cannot happen if you do not want to serve, care, feel, trust, be honest, accept the opportunity, and experience to learn how to use your brain, to change your mind. You have to be in the act to practice increasing your performance to reach this level of human, cognitive, and behavior processing.

Progressive Investing Model 33

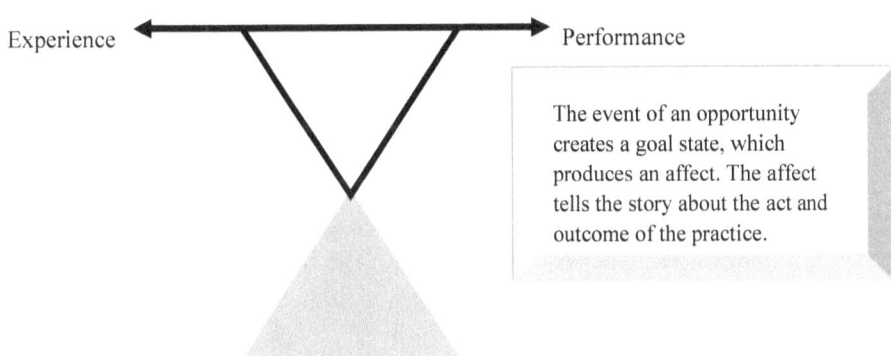

Session VIII

Stay in the game to play and to practice Progressive Investing.
If you learn how to use your home, school, and neighborhood for play and to practice, you will be able to focus the skills you need to experience your goals in the workplace. Every day you experience these places, you will learn how to use them to gather information, knowledge, and reflection from this work. As you begin to comprehend the work you are doing, the act to play or the act to practice this approach forms a learning plan.

The work of the brain helps you to receive, process, and respond to human, cognitive, and behavior processing. These processes are well known in the fields of human, cognitive, and behavior science. However, they are not known for their order and sequential effect on the brain, as receive, process, and respond cycles to being a human, cognitive, and behavior system. Hence, you are able to do the things we have described because of processing that goes on since you are a human being, a cognitive process, and a behavior system.

At each point, we have described how you move from being a simple processing system, to a complex processing system, and now to a superior processing system by learning how to use your brain to live, to learn, to think, and to respond. The problems you face at home, at school, in your neighborhood, and in your focus on the workplace are processing points, where contact with self, other people, and the environment has to be received, processed, and responded to a certain way.

You have to learn how to use the sense, feel, and focus cycle in accord with receive, process, and respond cycles to build the human, cognitive, and behavior processing skills you need to do this work from the inside out. This is what being human implies, this is what being cognitive implies, and this is what being a behavior processing system implies. All processing takes place in the brain.

If you want to make sense of being human, then talk to a person's brain. If you want to learn how to feel cognitive, then talk to a person's brain. If you want to focus your behavior, then talk to a person's brain. Your brain is now set up to sense, feel, and focus to receive, process, and respond to any signs of feedback. It takes these kinds of tools to learn how to accept criticism, advice, exchange points of view, and to focus a competent response.

You can become very adept at using your brain to respond to tension, stress, and pressure if you allow yourself to grow, mature, and develop these tools as open and flexible devises, from which to add or to build new skills. For example, we have talked about discipline and self-control. In this set up, acts of self-control come before signs of discipline, since the brain is the devise in use to manage and control the flow of any, and all feelings.

What does it feel like to be human? What does it feel like to be cognitive? What does it feel like to be a behavior? More importantly, you learn what it feels like to be a human systems thinker, which subsumes all three. Human, cognitive, and behavior processing changes a fixed response to an open and flexible response more set up to deal with change—live with change, learn with change, think with change, and respond with change.

You will use these skills to feel how being a competent human systems thinker feels.

The experience alone changes how you feel a sense of self, other people, and the environment to live, learn, think, and respond to all feelings. This is why we take the time to explain how you are here to serve your body to show true signs of care. If you have to trust somebody, that body is you. Do you have these skills? Be honest. Do you want the opportunity to set these skills up to be experienced?

Processing how you learn to use the brain to sense, feel, and focus, to receive, process, and respond to any, and all contact create competent paths of performance you can rely on through this practice. When we say behavior, you think with your brain. When we say cognitive, you learn with your brain. When we say human, you live with your brain. When we say human systems thinker, you live, learn, think, and respond to any contact in your home, school, neighborhood, and workplace networks with your brain.

A human systems thinker is processing sense data all the time—processing feelings all the time and processing lines of focus all the time. As a human systems thinker, you receive, process, and respond to processing contact to serve your brain with a sense of care for the state of your mind—all the time. Can you trust yourself the way you are? Do you want to live with the skill to free your mind? How do you feel this act to be skilled and competent? The experience—staying in the game to play and to practice Progressive Investing—creates skills that help you to act more informed.

We live to learn how to use our hands, eyes, and feet. We learn more through the use of our brain's hands, eyes, and feet to live. When you are in the act to learn how to live, your brain is processing the acts to sense, feel, and focus, receive, process, and respond. This is what we mean staying in the game to play and to practice Progressive Investing. Learn how to live each day to become more informed as a human systems thinker.

Progressive Investing Model 34

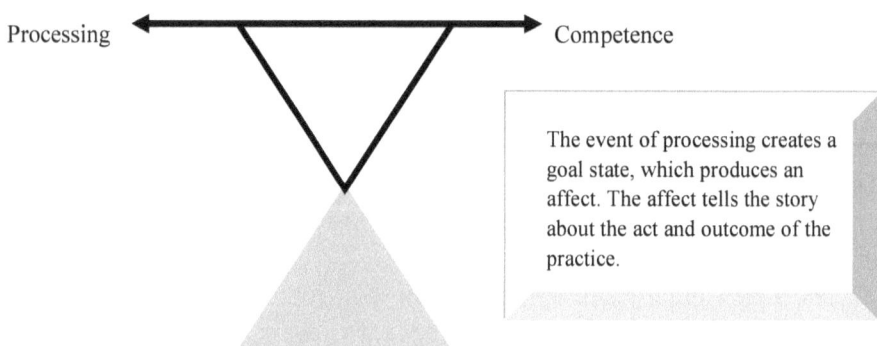

Session IX

Make the brain the home of your ideas to learn how to live.
In the last session, we may not have been clear enough about how to stay in the game to play and practice Progressive Investing. Progressive Investing is the act to live to learn why you have a brain to use in contact situations. Through the use of play or practice, you learn how to stay in the game. The home, school, neighborhood, and workplace networks are places you use for play and to practice experiencing contact with self, other people and the environment.

Your brain has to have a home you can trust. When you want to learn how to live, your brain supplies the ideas you need to move through the home, school, neighborhood and workplace networks. Your brain is used to decorate your home for Halloween, to create a science project, or to develop a lemonade stand through play and practice experiences. Your brain has to learn in the home you will use to move through the network. Your home needs to be a safe place to learn how to live.

Hence, make the brain the home of your ideas to learn how to live. This is why we use the words *to learn* before the words *to live*. You have grown up thinking that you act to live before you act to learn. This was because the body has been treated as a behavior that lives to respond in contrast to a brain that learns to live and think to respond. We hope that by asking you to make the brain the home of your ideas to learn how to live, you can make sense of this new human systems science.

We know how complex thinking about the state of your brain before you have been taught to think of your body as the home of your ideas can make you feel. This is why we have taken our time to share ideas about how backward a behavior can be when there is less thought about how it originates from any contact. For example, as you gaze at these words, what comes to mind first? The act to receive, or the behavior to receive the process of learning, or the process of living?

Our whole system of talking to your brain is based on common sense values and ideas about how you sense, feel, and focus to learn rather than to live, because learning is how you become creative in the act to survive. If we want you to grow, then we must say receive. If we want you to mature, then we must say process. If we want you to develop, then we must say respond. What we ask you to do is learn, live, and think through how you act and use talk.

How you think and use your brain to live implies an act to grow, in that a process is in use to help you mature as a learner, more so than as a liver. If you accept this idea, then the thought of being a human systems thinker implies that you have grown. If you have a grasp of this, then you can make sense of why you are viewed as being creative since your brain is acting to learn from the things you do with the hands, eyes, and feet to live. In other words, the soundness of your ideas flow from how well you are set up to serve—to make the brain the home of your ideas; to learn and to live; and to grow, mature, and develop.

Being set up to think helps you to develop your ideas, which in the process, makes you more creative, as you learn over time, how to live with the person you have become. This is how we use talk. This is how we to talk to the brain. This is how we talk to your brain to get you to think about the ideas we have to share in a way that allows you participate in a creative science to help you learn how to live and respond. Your brain is the home of new processes for changing the way you use the world to sense, feel, and focus, receive, process, and respond, and grow, mature, and develop ways to play and to practice Progressive Investing.

Living to learn how to live in a home, living to learn how to think in a school, living to learn how to think in a neighborhood, and living to learn how to respond in a workplace are values of Progressive Investing. When you place your brain, in the game, you live to learn how to use these places for play and to practice the act to learn, live, think, and respond through your brain's work. You think through your use of these places to create strong ideas.

Progressive Investing Model 35

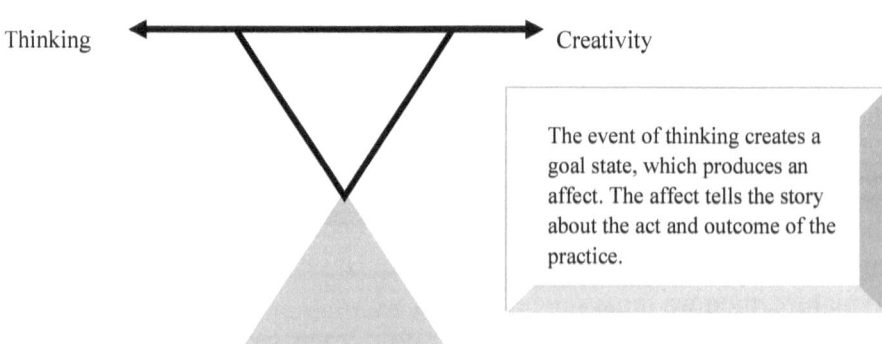

Session X

Is there a choice, or do you just say you have one?

When you have well-built ideas from the way you choose to use your brain, then you can say you have one. For example, you may be having a hard time with the people you live with, but you work hard not to let these problems upset the path you have chosen. You may be having a hard time with the people you learn with, but you work hard not to let these problems upset the path you have chosen. You may be having a hard time with the people you think with, but you work hard not to let these problems upset the path you have chosen.

If someone says he or she trust you, what does it mean? In other words, he or she is trying to talk to your brain. Do you mean to say and do the things you say, or do you just say the things you want to say. Is there a choice, or do you just say you have one? If you mean what you say, then you are making a choice. When someone trusts you, he or she feels a sense of who you are. You need a strong sense of self to stay on the path you have chosen.

Think about this: Your brain is your body to learn how to feel and focus. Is there a choice, or do you just say you have one? To help structure your sense of self, as a human systems thinker, you have to decide which one you trust most. For example, are you mental, or are you more physical? Are you physical and mental, or are you mental and physical? At this point, you ought to have a sense of why; we want you to think about this. Do you choose to lead with your body or do you choose to lead with your brain? You have to want to stand by the choices you make.

If you choose to lead with your body, then how you feel and focus is secondary to how you look and behave. This is because you judge things based on how they appear to you in the physical sense and how they respond to you in the physical sense. You have been taught if you cannot see it, then it must not exist. Hence, you deny you have a brain to use, or you choose to live as if you do not know for sure you have one. You have a brain in your body. You can ignore there is a choice, or do you just say you have one? Your strength and resolve comes from the path you choose.

If you say you have a brain to use, then you can learn the value of your brain the balance of your natural life. This means, you learn to live each day to become more informed from the choices you make to the decisions you make to form the ability to sense, feel, and focus how you judge things by their content rather than by their physicality. Instead of just being a physical human system, you choose to be a mental, physical, emotional, and social human systems thinker. You look to learn from the things you do and say since the things you say and do come from your feelings of being led.

This is how you think: Your body is guided by the brain's power to analyze the things you say in contrast to the things you can truly do. Furthermore, the brain has the capacity to synthesize, pull your experiences together, and sift through them with focused hands, eyes, and feet. There is a choice. You have to decide what you are. Are you an unknown object of behavior, or are you a human systems thinker? What type of brain are you? We want to know, but you need to know more. You are the judge, can we trust you? Will you judge our talk by the content of the words we use to reach for your brain, or will you ignore that these words are meant for your brain?

You have one life to live. Are you ready and willing to serve—observe, learn, listen, help, and lead your human system in new ways of learning? Think about this: Is your brain ready to lead you right now? Right now, at this very moment, do you trust your brain's capacity to lead you? Is there a choice or do you just say you have one? Which do you learn with, your brain or your body? Are you here to serve? How do you choose the right path to take?

Progressive Investing Model 36

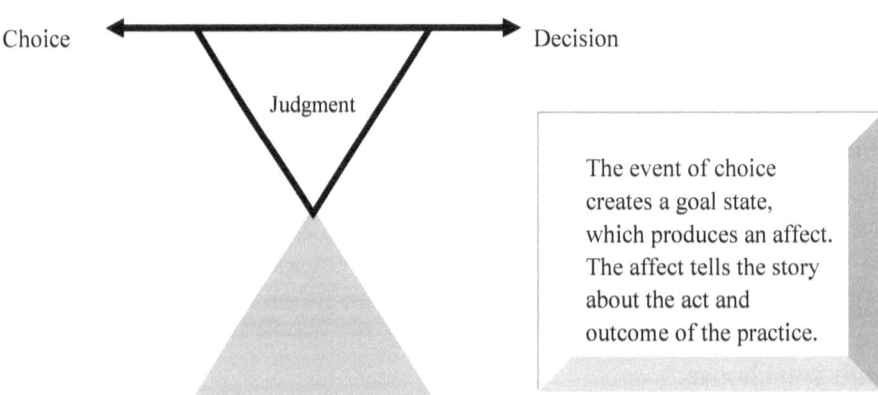

Session XI

How long have you been living on hope? Are you on the right path? A human systems thinker learns how to use them all to say he or she has one. If you cannot feel the sense of doing this work to learn how to live, where do you think your brain will learn this? You may feel the need not to want to focus on the words we use to ask you. Do you care? The choice is yours to resist. If you feel what we are trying to say, you are learning.

Now that you have a sense of how we talk to the brain, is your brain open? How do you assess your brain's open path? Are you here to help or are just here waiting to be helped? How long have you been living on hope? You are here to learn how to serve time to help move past the fears of not knowing how to help. You can ask for help, but you cannot ask for hope. Hope lives in a poor state of mind or a mind that is coming to help. You want to ask for help, not hope. You want to move your brain's body to learn how to search for help.

Your brain is set up to help you learn how to help. Your body is set up to hope you find the help you need. If you are waiting for help, then you are dangling on hope, when you can use your brain to find the help you need. The next step is to learn how to move beyond living on hope to learning how to find the right helper's help. These words are meant to help you think. What is the value of a sense of help in contrast to a sense of hope. Life comes at you fast, hard, and steady. You want help, you have to ask for it, but if you want hope, you have to find it through your act to care to feel your brain at work.

Hope is not something you can measure, like help. Hope may seem like trust, but points to a dangling sense of despair. Now, a sign of despair, you can assess. The brain may appear lifeless, where the body seems tense and laden with signs of tension. Now think of help. Help is a quality. A quality is a feeling of value or high regard. A feeling of something impressive can only be valued, if the brain is open to learning how to feel this form of life. Hence, you have to be ready to serve, help the helper help you feel them. You can assess the quality of another person based on how they use words to talk to your brain in contrast with your body.

We say there has to be a balance between feeling and looking, focusing and behaving. If a person wants you to help them help you, then he or she will think about how you feel in contrast to how you look; and how you focus, in contrast to how you behave. We do this to assess the state of your brain in relation to the state of your body to observe, learn, listen, help, and lead you to accept the help we have to offer. You can measure this form of helping by the words we use to talk to your brain to help you learn how to help. How long have you not known how to help? We can assess the feelings of resistance flowing through your hand movement, how you hold your eyes, and the way you move your feet.

If you want to live on hope, then you live to uphold a physical state of bad feelings. People look at you, feel your signs of despair, and want you to feel better, but you resist their contact. You do not want to see them, you do not want feel them, and you do not want to learn how to help them help you. How long have you been living on hope? Do you want to help lift yourself up to a higher quality of life? If so, then you must learn who you are when you feel hurt and what you are when you feel hurt. When you feel hurt, you look as if you are hurt and you behave as if you are hurt. If you feel hurt, but try to look as if you are not hurt, then your feelings of hurt begin to focus how you behave.

When you are looking for help, you are learning how to clear your mind to feel signs of care. You can assess signs of care by the way a helper seeks contact, interacts, and tries to cooperate with you, even through high signs of resistance. The words he or she uses are meant to calm you down, so you can observe them, learn with them, listen to them, help them, and lead them to help you help yourself. You measure the quality of their state of mind through their signs of care. When the signs are there and you resist care, then you have to accept how long you have been living on hope free of care.

Progressive Investing Model 37

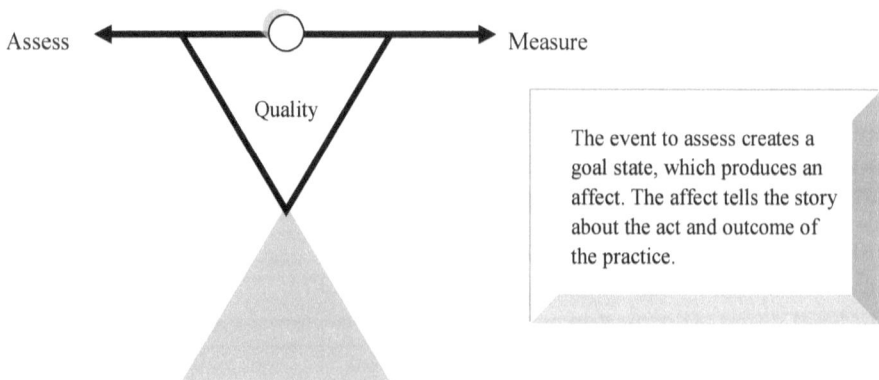

Session XII

What do these words mean—*observe, learn, listen, help,* and *lead*? Trust me, you have not felt like this in quite a while. This is how you feel when the words are for you. You learn from the way the words tests your sense of care. This is why; you can feel the act to focus, how you show signs of care. You focus to make sense of the words to feel them with your brain. This is how your brain learns to read with a focused sense of care for the words.

What do these words mean; observe, learn, listen, help, and lead? In our book *Learning How to Live*, there is a field of perception. We study how you explore, experience, learn, and reflect through a lack of knowledge of the external world and the people in it. This is to say, we look at both how knowledge is constructed and how knowledge of the self forms a skill. Knowledge of the mind, knowledge of the body, and knowledge of the universe is required to say and to do the things you say to learn how to live.

When you live in search of an awareness of your role, you have to be able to observe, learn, listen, help, and lead the way. Otherwise, when the roles you search for is within your grasp, how will you know it? For example, when you have found the right helper, you will know it, because of your act to observe them, learn from them, listen to them, help them, and lead them to prove their signs of care are for real. In any situation with a helper, you have one path to use to improve your chances of survival. Look him or her in the eyes to feel them and focus on how he or she behaves when you release a smile at them to learn from them.

From this point forward, the relationship forms a state of awareness. Learning how to live means learning how to form relationships that you can trust through your good feelings being held up by the other persons. If you show signs of care and he or she show signs of care in the exchange of contact, then both parties must hold up these two states. What do these words mean—*observe, learn, listen, help,* and *lead*? These tools are used to assess and to measure the quality of his or her signs of care. This is why you have to learn how to live to accept help. To move beyond a state of living with your pain from the outside in, you move to a state of learning through the act to feel the pain from the inside out, to survive.

You have to act as if you want to learn how to live to move beyond simple states of survival. You act to learn how to feel the painful state of your brain and free your brain to observe, learn, listen, help, and lead you to assess the people in your field of perception through signs of care. Your smile changes the way your face feels, looks, seems focused, and your body seems to behave. People respond to the things they see in you, from the outside to the inside. When you are mental, this is help, to learn from this act, to learn how to live.

You have to set your brain up to feel your body in the world of contact to connect the exposure, to the experience of learning how you reflect. When you reflect, you call to mind both the good and the bad feelings. The choice is yours to learn. There are consequences. How long have you been living on hope? There is a choice or do you just say you have one? What do these words mean—*observe, learn, listen, help,* and *lead*? Are you here to serve, or are you still waiting on hope to choose you?

Who is coming to help you? Is your brain in your body to help? How do you create a body full of ideas to learn how to live in a safe home? How long can you live free of your brain? If you use your brain to learn how to live through good and bad feelings, you are learning to use your body. The choice is yours to learn. You know you have a body to use since you can see it. Now, you know you have a brain to use since you can feel it.

Progressive Investing Model 38

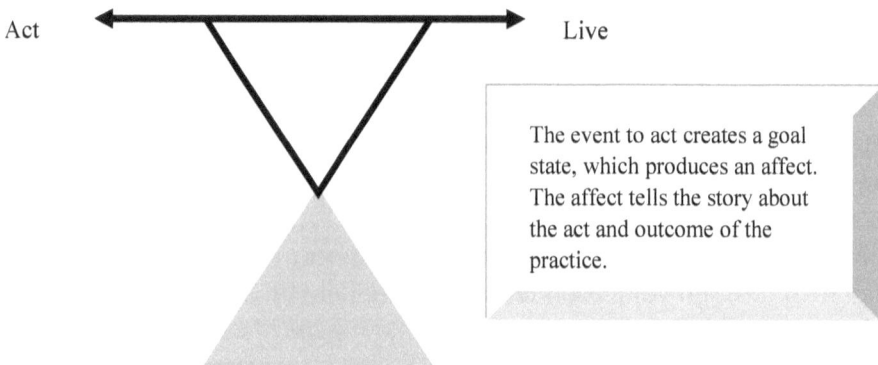

Act

Live

The event to act creates a goal state, which produces an affect. The affect tells the story about the act and outcome of the practice.

Session XIII

Know what the goal is in the aim. Can you feel you brain? This is science. What do these words mean—*observe, learn, listen, help,* and *lead*. These words call for you to free your brain. Let your body work with you, but not free of your brain. You can seek to learn how to live with your brain's body, this act of science. The choice is yours to learn.

Know what the goal is in the aim. You take on the act to learn how to live from an event state, to a goal state, with the aim in mind. Learn from where you are: how to feel, look, focus, and behave in the end. The goal is in your aim for results from the act to learn how to live. You feel your brain, you look at your brain, you focus your brain, and you behave with your brain showing clear signs of care through your action.

The goal is in the aim to learn through your awareness of how you feel, how you look, how you focus, and how you behave with care. You are here to help yourself learn how to feel good, look good, focus on being good, and to behave with care. Is this a goal, to live with a sense of care for self, and for others? The choice is yours to learn. When you move from the act to aim how you live and use your brain to think, you learn. Awareness is the result when you take the time to set up a goal in the aim of your act to serve your brain's body. You learn how to live with a goal that helps you to aim how you use your hands, eyes, and feet.

Now you know people can see you, because you want them to learn how to live with you, to learn how to live with them. You are learning how to be seen from the inside out to feel, look, focus, and behave on the outside based on your feelings of reflection. When you can feel the action, when you can look at yourself learning, and when you can focus your awareness, people see how you can behave and want to learn more from you. The choice is yours to learn. Do you want the helper's help? The tools we have talked about thus far, have they been helpful to you? This is the time to choose how you want to learn from the balance of your natural life.

We only have the time we have right now. If you choose to resist, then you will lose this opportunity to learn. Is there a choice, or do you just say you have one? How long have you been living on hope? What do these words mean—*observe, learn, listen, help,* and *lead*? Do you know what the goal is, and are you in the aim? In other words, do you know your role, and are you in the game of life to win? The choice is yours to learn. Decide now how you want to live. Do you want to be rich, middle class, or poor?

Progressive Investing Model 39

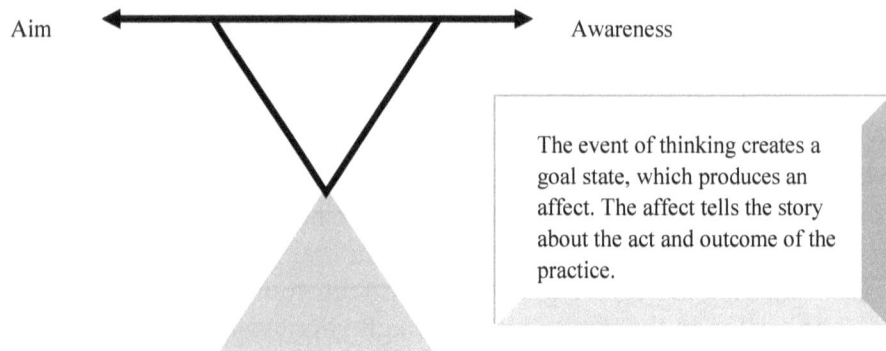

Session XIV

Looking back, can you feel the help? These words are not meant to scare you. These words are meant to help lead you to think. What do you want to be? Each step you make helps you to adjust to these words and what they may mean to you as a science of learning how to live with self and the people you meet to learn how to compete in your home, school, neighborhood, and workplace environments. Are you there yet? Can you help?

Looking back, can you feel the help? Your brain's body is an act of science, all the time. As you have read the words we have set up for you to consume, you have an idea in your brain about me. You sense this: How I think, how I think we learn, how I think we feel contact, and how I think we have to learn how to live. The things I believe count in life; these are the people who learn how to care and exchange contact to help serve the people.

You might want to know who the people are. Look back, you can feel the help. Focus on this. Plan how you want to learn how to live and whom do you want to do that with. We have to make sense of people since our plan includes learning and living through the experiences of people. As you reflect on these words to learn how to help them serve your brain's body, you can feel the help. I am here to help you reflect.

Look back: This is not about me. This is how we set up your ability to use these words to reflect. Can you feel these words seeking to make contact with your brain to share signs of care that may help other people learn how to try?

Learning how to help is an act to help other people look back to feel the helper trying to help him or her think to help. Being able to learn from where you are means these words can move you to clear your mind or look back to free your brain's body to help. What do words mean to a learner trying, to a student studying, or to a scholar researching ways to move out of a rut? At the beginning, they are just words. At the apprentice level training and at the academic level, this is knowledge on how to care for the brain's body. As a child learns how to learn, as a youth grows to be a learner, as a young adult matures to be a student, and as an adult develops to be a scholar, he or she is learning how to live through the use of talk and words.

Looking back, can you feel me? Look back, can you feel us? Look back, can you feel the help? As a child, a youth, a young adult, or as an adult you want to use the brain's body to observe talk, to learn to receive words, to listen carefully, to help the helper, and to lead the process of learning how to live with self, other people, and the environment.

Yes, you can look back over these words. The choice is yours to learn. How you learn to use your brain's body is by learning how to learn from other people like me. To understand why, we use special words to talk to your brains body as signs of care. Sometimes, care is seen as a weakness, so we use straight talk. Think about the things you say and do.

Your brain has to be open to these words; know what the goal is in the aim. This work is about you, not me. Now, looking back, can you feel us? Looking back, you can feel you at work with me. Let us be frank with each other. Can we help? Can you learn to accept the helper's help? Learn from these five words: *help the helper help you*. Think from these six words: *help the helper help you help*. The choice is yours to learn. Learning can be simple, complex, and advanced. To look back means to reflect on the experience of learning with us.

What are you doing to help your brain's body, learn how to live with you? You looking back, you feel what can be done to help in contrast to what you will do to help. If you resist, the choice is yours. If you want the help, the choice is yours. Thinking forms the science of how good choices are made. When you look back, you have made a choice to think. When you think and you know it, you look back. Yes, you do, to learn how to move with care, you look back.

Progressive Investing Model 40

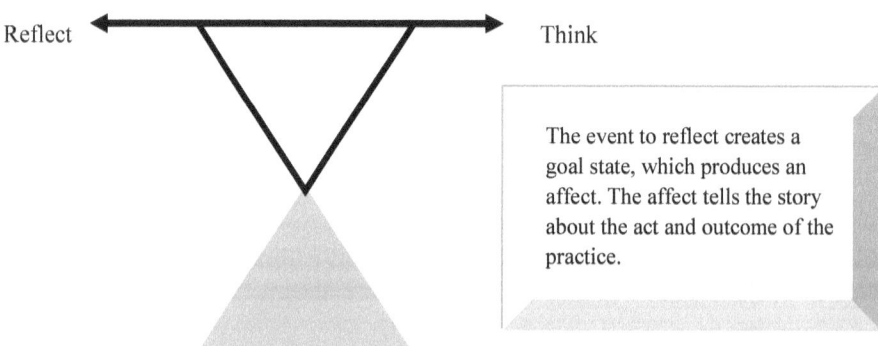

Session XV

You can do the brainwork, and we can set it up. It may start with me, but it will end with you. Through acts of science, we will teach you how to learn with your brain's body, to grow as a human being, to mature as a human system, and to develop as a human systems thinker. The choice is yours to think. The brain's body has to learn how to make the choice. Free your brain's body to sense, feel, and focus as a human systems science.

You can do the brainwork, and we can set it up. When you are hurt, it hurts to care. When you do not feel as if you care, you think bad thoughts. To heal, you have to learn how to feel. You want to feel healthy. Open your mind to read the words we have shared.

This time, take a closer look at the words. Look for how the words may relate to you and how you feel right now. As a service to yourself, look back and review our first session. What do we mean by serve time?

These words mean more to you. When you look at these words through an open mind, what they may mean to me is not the same as how they may relate to you. When you think like this, care feels like you are weight lifting. You have to sense from the people around you how the work may pay off in the end. You need to trust the reason why you do this work, you want to help yourself, and you need to feel healthier.

When you are being honest with yourself, these words help you to recover good thoughts from bad thoughts to feel a sense of focus. The opportunity to feel you change from a lack of care, to the experience of good feelings of care, can make processing these words a healthier act of science. You have to care about your brain's body to create a state of discipline through the use of self-control as a skill, now that you have a sense of what the self feels like to be thinking.

The choice is yours to learn how to use your brain to feel healthier signs of care. This is a doable science. You look to study how you feel as an act of science. If you aim your brain's body to learn, you think of ways to live with care. When you can think of ways to live with care, your brain's body is safe to learn. Care costs: you have to be able to look back. This is why it hurts to live with bad feelings. Learn how to reflect to think with the skill to dump signs of bad health.

This plan is fresh and easy to read. Each step is plain, straight talk. The words are organized, the learning cycles are structured, and the Progressive Investing models help you to sense, feel, and focus each event, goal, and possible affect on ways to solve problems. You want to be healthy. You can do the brainwork, and we can set it up. How do these words run through you? You are testing this science. How do you feel the acts of science? Do you use care to receive the words, to process the words, and respond to the words?

Progressive Investing Model 41

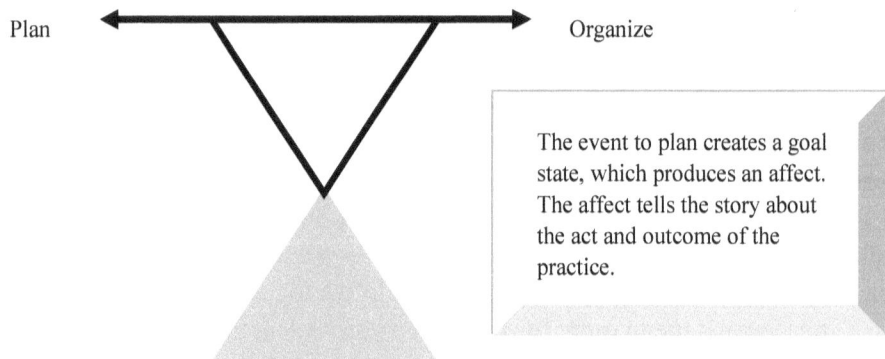

Closing Session

Save this work to heal. Is there any doubt you have a brain to use? How do you know when your brain is in use? We had to ask those two questions since we are here to help. We think this is a new science. Do you know how to sense, feel, and focus words when you are hurt? How do you know when your brain's body is set to learn? We look at how you receive words to learn from your act to participate in the use of these words. Does this work have value to you?

If you are hurt and you will not talk about it, how do words flow through you? Words that come in touch with your brain's body need a path to move through you. You can test this path. Your brain's body is used to sense, feel, and focus on a choice. There are paths to receive, to process, and to respond to the choices you make. How words move through the act to sense in contrast to the process to feel the act is influenced by your state of mind at the time.

For example, when you choose to live in a home, you try to make sense of it, to feel it, and to focus the act to use it in the process of living in it. If your feelings are hurt in any way, this may affect how you try to experience the things that relate to you and your home. When you do not want words to flow through your brain's body, you resist any need to receive sense data, which relates to the way your home feels. This means your sense of feel is resistant, and hence, the act and process to receive a sense of balance to focus may not respond.

Save this work to heal. When you hurt, you may just shut down. We are here to help. Do you know why you are here? Your brain's body has to work hard to help you learn. Not to live but to help you learn how to live. You live in the body of your brain. When you cut the path off to your feelings, you cut the path off to process your act to learn. You give away your chance to learn how to process these feelings. The choice is yours to learn; you want to learn how to live in your brain's body.

You resist the choice, and you may have a problem in how you try to learn to live. When you look at how you live in a home, learn in a school, think in a neighborhood, or focus in the workplace, you are in the act or process to try to experience, or not to try to experience them. If you do not have a sense of how you feel in these places, then how can you focus to learn to use them? One place influences the next in terms of how events and goals affect your brain's body. If you learn how to live in a home, then you can learn how to learn in them all the whole network.

When you try to live in a home, you have to learn how to interact, cooperate, and participate in the act to share it with other people. If you can learn how to act to feel the people in your home, then you can learn how to interact with them. To work with your brain's body, you have to cooperate in the act to learn from others how to move through them. When you are hurt, you hurt on the inside. When your home is hard to live in, your home is hard to live in on the inside. Care costs.

You act to sense, to process the act to feel the affect, in the processing of the act to focus the contact in a home. You learn from the act how to sense, feel, and focus to set up the way you interact in the role to participate. You live in a home to learn how to interact in the event to cooperate with the goal to process your act. In the role to participate, your act to focus will have an affect on the path you choose to cooperate. The senses are used in the aim of the act to learn how the role feels to be focused with care.

The act to serve is the event that transfers the path you choose to use in play or practice to reach a state of self-control through the process of thought. The act to sense, feel, and focus the processing of the act, transfers how it feels to have feelings in the path of self-control. You choose to feel these feelings out, to learn how to interact in the aim of the act to cooperate in the use of role-play or the practice of participating. Save this work to heal.

Using the sense, feel, and focus cycle in the act to process contact, you cooperate through focused thought to set up your choice to learn. Your mind is open to the internal event to receive sense data in the brain's body to sift through it. Your state of mind is open to the goal to respond in the process of thought with action to learn how to live with self and other people. The environment—home, school, neighborhood, and workplace—is used through play or through practice to learn how to live with a since of discipline and self-control to improve how you feel them.

Progressive Investing Model 42

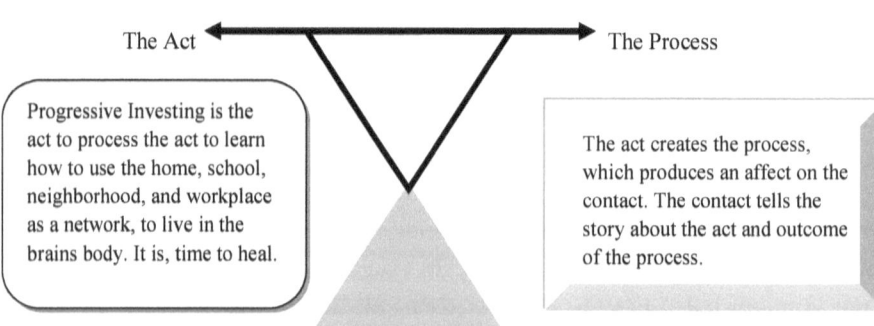

Workshop Notes:

Now that you are experiencing human systems thinking, tell us how you feel.

Tell us, what is in the experience of human systems thinking?

Select one (learner, student, researcher, helper, and participant) role and tell us how you experience the effect of being a participator.

Comments:

Submit your request for additional sessions.

☐ I would like to set up a date, time, and location where my team and I can receive more Progressive Investment Training.

My name is _____
My title is _____
My e-mail is _____
My mailing address is _____
City _____
State _____ ZIP _____
My preferred dates are _____
My preferred times are _____
My preferred location is _____

☐ On-site
☐ Off-site

FOR A BUSINESS REPLY

Use the self-addressed envelope that is enclosed with this guide. The envelop should be mailed to the following:

Progressive Investment Group
Progressive Investment Training
ATTN: Progressive Investing Institute of Focused Learning
P.O. Box 278363
Sacramento CA95827-8363

Chapter 7

A Focused Learning: Human Systems Research Perspective

The purpose of focused learning is to show the participant how to set their brain up for contact. This is why we want the learner to learn how to use the brain and body as a human system. The learner has to be set up for the role to explore, experience, and experiment with the act to live, which is in the physical act to be a sign of life. Hence, the learner is being set up for the process to confirm, care, connect, and learn from the contact, which is in the affect on the goal to act as a participant. These two steps include the act to observe, which allows us to set the learner up to transmit the goal to send and receive feelings that relate to a sense, feel, and focus cycle of learning. In other words, the learner learns how to manage a flow of energy.

In focused learning, when the brain and body works in concert, he or she is called a human system to account for their mastery over the brain's body. The brain's body is a human system when the user learns how to control the flow of energy in his or her sense, feel, and focus cycle. In other words, the sense, feel, and focus cycle is set up to help the learner learn how to manage the flow of energy. For instance, the learner learns that energy flows through the body, but when the brain is the lead, the energy becomes more controllable.

The energy that flows through the body is real and tangible. The learner learns that when the brain is in the lead, the body becomes a real sign of focus. For this reason, we say the learner's brain is in the aim to live, which means that he or she is set to learn with an aim in sight. If we say the learner's body is set to live, then we are saying that he or she is in the act to focus the brain. In other words, the body serves as a physical sign of life or proof of focus, in that, to show a sense of the brain, he or she has to transfer words and talk, which form learned states to feel from experience.

Progressive Investing Model 43

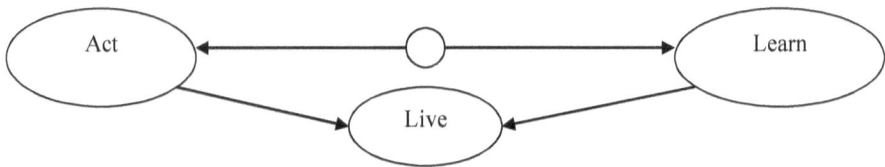

A Progressive Investing Model seeks to show in real life how a person may use the brain in the act to gather a sense of time, learn to feel space and time, and live to aim the sense, feel, and focus cycle across energy, space, and time.

The act is in the goal to learn, which produces an affect on how an actor chooses to live. The aim to act focuses the goal to learn from the act to sense and feel the affect, which is to live within the aim, as a state of mind.

Focus is in the act to learn to use the brain's body to think as a learner's physical and mental skill to improve how he or she lives as a human system. We propose that all physical and mental skills are a result of learning and thinking through stages of focus. Thus, to learn or to think, a learner must be able to focus the body to aim the mind as one complete act. Hence, an act to focus turns out to be a formula for the use of talk to hear, listen, transfer, interpret, think, and react to words and live to focus the brain.

Progressive Investing models are used throughout this book as samples of how we use simple words to talk about complex problems. The words we use to talk about the problems we seek to solve help to create special words from the practice of looking at an event, affect, or goal as a cause, relationship, or effect between these points of reference. We explain this process in *Education and Science* as the analysis, assessment, and synthesis of a problem through a focus on the person as a human system.

OVERVIEW

Experiencing the Experience of Practice

An experience is something that can be felt in the act to live and repeated as an act to learn through acts to reflect. The act becomes the actor's role to learn how to perform. When the actor is exposed to the event, the actor experiences the goal in the role to experiment. Hence, the actor's goal is to practice as a learner from the effect when contact is aimed in a situation, or process to focus the brain's body on the act to learn. Hence, to focus and to learn how to use the brain is an experience of the senses.

The sense experiences of an actor in home, school, neighborhood, and workplace events are cause-and-effect situations. The learner will have to learn how to move through them with a sense of energy that relates to the way emotion may feel, as it flows in the brain's body, in search of a focused state to think. In the event stage, the actor has to feel the energy and emotion through separate experiences as a force and as a feeling. In the aim to reach for the goal to learn from the experience, the actor's brain's body will search for a line of focus to connect with the learner's role to live through the contact.

This is the point in the experience where the actor learns how to feel energy to capture emotion through the use of the senses to focus states of inner growth and self-actualization. We are not attempting to be complex; we are, however, attempting to set the stage for the experience of how to focus a learner's act. The learner seeks to feel the first act to live and the second act to learn how to live through the experience of both, feelings of energy and feelings of emotion. Then there is the third act to think, which is yet another experience, but is the vital experience that sets free the actor's capacity to reflect or look back at past experiences. In human systems research, we look for evidence to explain how the learner observed the experience, learned from the experience, listened for the experience, helped to form the experience, and took the lead role in the experience of the practice.

If you are an actor in the role of a coach, teacher, mentor, counselor, or participant, you want to experience this effect to learn how to focus the learner's brain's body to practice the experience of being the experience too practice. The actor has to make sense of these feelings. The real goal is to live through the experience of energy and emotion as separate states of learning how to focus the brain's body. You want the participant to release emotion, seek inner signs of growth, and to self-actualize in the process of energy to live.

Progressive Investing Model 52 conveys how the act to sense is the event in the actor's goal to feel the line of focus, which has an affect on the practice of the event to sense, feel, and focus the brain's body.

Progressive Investing Model 44

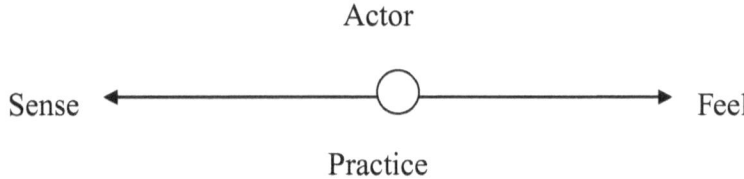

Use the senses to feel for the effect of the practice as an actor. The body is set up to live through the contact in search of ways to feel the contact to learn what causes it. Hence, the brain is in the act to come to think as the practice moves the actor to sense and feel. The acts to focus relates to the interaction, which is the bridge between the events, goal, and affect states of mind. The choice to cooperate or to resist the need to respond concludes the sense, feel, and focus cycle. The actor participates or fails to participate in the acts to live, learn, think, and respond to the contact based on these values.

The brain's body is set up. The experience of the event is to sense and feel the effect in any contact, which will carry a dual or common affect on the actor or learner relationship. To learn more, review the focused learning sessions that follow. These experiences flow from more than fifteen years of Progressive Investing. Progressive Investing is the act to live each day to become more informed. This is a lead up to "A Focused Learning: Human Systems Research Perspective."

Session One

Coaching

In focused learning, coaching is the experience of knowing how to use contact to study the act to live through it, learn through it, think through it, or respond through it with the aim to participate. The experience of setting a participant up comes from the coach's sense of how to lead him or her to aim the body to accept information using the brain, as the place to aim knowledge to create awareness.

1. For example, in the experience of practice, the coach experiences focused learning to become the experience of a learner's focus, the skill of which produces one line of discipline in the form of self-awareness and self-control and to focus overall feelings and to organize the senses.
2. The ability to gain a sense of the act, which has to include focus, produces the interpretation of the interaction as a technique for reading both body and brain events.
3. Coaches have to know what it feels like to have a sense of focus in order to show a participant the steps that lead to a focused state of mind.
4. Hence, the experience the participant receives is authentic. The learner learns from the exchange of contact how to feel being motivated through the tasks of sensing, feeling, and focusing the act to perform.

Progressive Investing Model 45

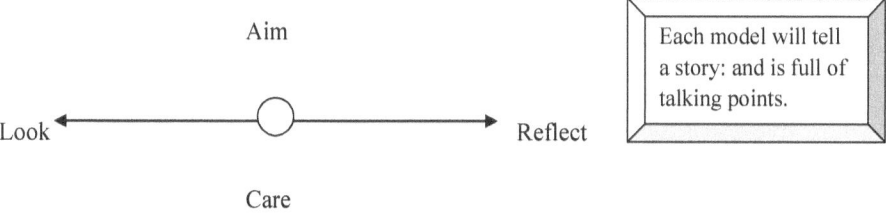

The act to look guides the process to reflect to show signs of care for the need to help aim the brain. The brain is set up to receive sense data to help the actor learn how to use it. Hence, to reflect means to process energy and feelings through the aim to think with care for the affect. In this case, the affect is what the actor responds through as the brain's body looks to reflect or fails to aim signs of care.

A group of words will relate to the way the human system develops intellectual functions for sensing, feeling, and focusing the flow of information. The Progressive Investing Model tells the actor or learner to look at the problem or issue to set up the live, learn, think, and respond cycle to improve the ability to observe the contact, until interaction reaches an optimal level of performance.

Teaching

In focused learning, teaching is the experience of learning how to use contact to study the act to sense, feel, and focus the aim to care. The experience has to be transferred to set the participant up to organize and form a sense of the act to learn with a body and brain that cares to be connected to the practice of feeling like a learner.

1. For example, the act to teach begins with how it feels to learn how to experience care in the aim to practice the art of teaching as two separate and distinct experiences, within the process to learn from the exchange of contact with the participant to aim signs of care as a skill.
2. The teacher experiences an effect, which produces an affect, which creates a set of focused skills, which transfers the practice as the base from which caring emerges.
3. Teachers have to know how to feel contact, adjust to it, which means to interact with it, which in turn means to cooperate with the need to focus it.
4. To know how to help the participant accept information, the teacher has to know how to lead the participant to experience care from personal experience, through receive, process, and respond cycles of learning.

Progressive Investing Model 46

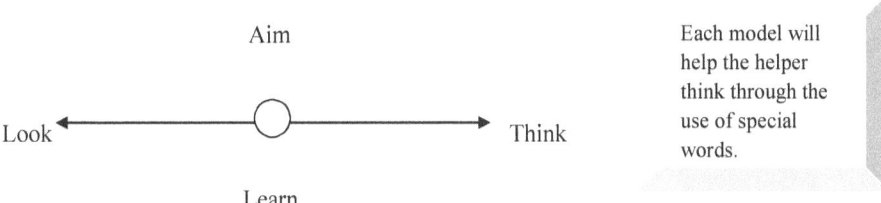

The act to look activates the line of focus to receive the process to think in the aim to learn how to respond. The brain is set up to sense and feel the contact to focus receive, process, and respond cycles as pathways to learn from the aim to think. To think now means to look for a line of focus that can be gained through the interaction and connection of these two cycles of physical and mental processing.

A group of words will connect the physical, mental, emotional, social, and experiential changes that occur in a learner's human systems receive, process, and respond cycle as the basis of the internal effect for the control of feelings. The Progressive Investing Model tells the actor or learner to feel the aim within the goal of thought that focuses the human system to learn from the physical force of an affect in order to arouse the aim: to think to learn.

Mentoring

In focused learning, mentoring is the experience of the act to help the helpee help the helper lead him or her to practice and to accept signs of care.

1. For example, a mentor experiences the act of being a helper to focus the aim to help build the learner's sense of self and the home, school, neighborhood, and workplace network as places to practice.
2. Mentoring is the experience of setting the learner up to experience contact with the people who live, learn, think, and respond in these places to practice.
3. A mentor has to feel the help of others in the aim to lead the learner to learn how to move through these places to practice: to

learn how to help him or her care for the way they grow, mature, and develop through the help of other helpers.

Progressive Investing Model 47

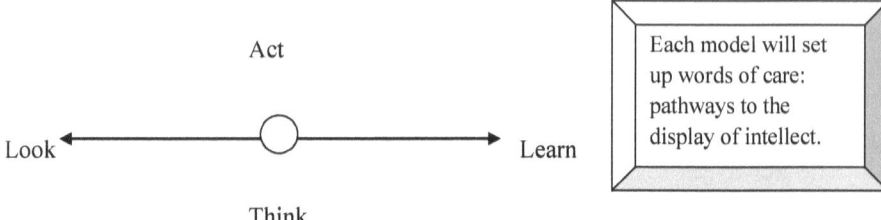

The aim to look forms the need to learn from the exchange of contact how to think through the act to live. The body is set up to receive sense data to look at how the brain learns. The process of the sense data leads to an affect, which is to think through the aim to learn. Hence, the act produces an effect to aim the goal to learn through social and emotional events. To look relates to the need to learn how to move through home, school, and neighborhood environments to learn how to look, think, and act.

A group of words will encourage the use of intellect, which opens the pathways to experience and reasoning; the general goal of any human system is to comprehend the practice. The Progressive Investing Model tells the actor or learner to become the experience to produce an affect, which helps develop the skill to sense, feel, and focus human, cognitive, and behavior goals.

Counseling

In focused learning, counseling is the experience of talk aimed at the learner's brain as a practice of care to observe, learn, listen, help, and lead him or her to study the words that are used to feel and focus contact.

1. For example, the interpersonal experience of the live, learn, think, and respond cycle is a reflective process by which the actor draws from the personal act to cope and move beyond problems of living.

2. For example, the counselor may make use of the sense, feel, and focus cycle to discuss how words are used to cause the learner to experience emotion as a reaction to being in contact situations.
3. Talk is used to study special problems that include learning, behavior, and conduct relative to focused learning steps to improve the flow of information and process goals.

Progressive Investing Model 48

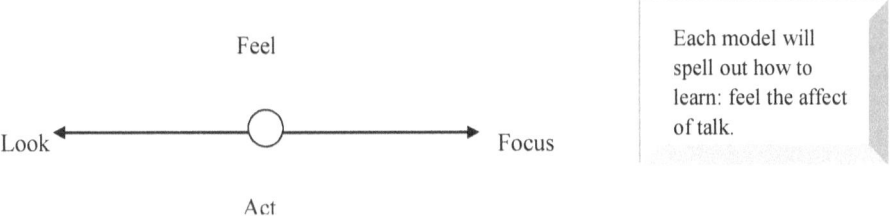

To look is to focus the act to feel as the affect of the practice. The brain is either open to the flow of energy or resistant to the feeling of contact. Words have to be received—accepted through the act to sense and feel them; to focus the process to live, learn, think, and respond through them as signs of care. To look means to observe, to focus means to listen, to feel means to learn, to practice means to help, and to act means to lead.

A group of words will act as a guide to enhance the aim to learn through personal problems by aligning the experiences they mean to address. The Progressive Investing Model tells the actor or learner to think about the goal from several points of view that are affected by the problem to achieve the special focus of learners.

Team Development

In focused learning, participating is the part of the experience that includes the act to take action in relation to contact, interaction, cooperation with the intent to focus the effect of feeling, each of these aims as a team.

1. For example, the team participates from the perspective of how he or she needs to live each day to become more informed.
2. The act to live sets up the aim to accept the experience of being active in the role to focus.

3. The team experiences how it feels to feel self, the environment, and other people in the study of home, school, neighborhood, and workplace functions to practice focused learning as a learner or in the tasks to be a leader.

Progressive Investing Model 49

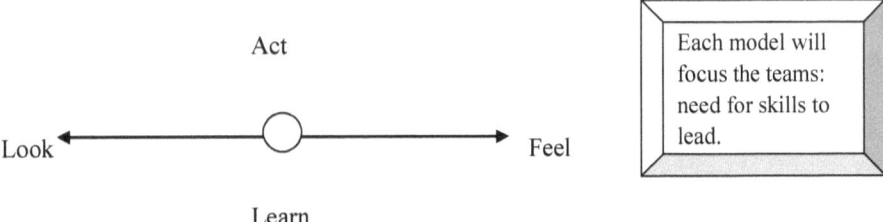

The need to look forms the path to feel how the brain learns the act to respond. The body is set up to receive the need to look at the problem from the inside out, to feel the process to learn form as the experience of learning unfolds. To respond to the need to look, how it feels to learn has to be recognized in the act live through, learn through, think through, and hence, feel. Every step—to sense, feel, and focus; to receive, process, and respond; and to act to sense and feel contact, interaction, cooperation, and participation—is aimed as the need to live, learn, think, and respond forms each path.

A group of words on purpose or by accident will lead to facts, skills, values, and ideas through the act to participate in the aim of the human system. The Progressive Investing Model tells the actor or learner to feel the path where passion often has to be toned down, in that the role affects the team's sense of the act to learn how to practice.

Session 2

Problems of Focused Learning

Given the growing body of knowledge about the influence of the sense and receive data cycle, it seems that the aim to focus the brain, body, and senses are affected by a person's lack of awareness for how the human system works. Focus is a process engaged through the ability and desire to aim the senses to be open to acts to feel an event, object, or thing in a state to learn from the experience. When this does not happen, is it because of ignorance or a resistance to the aim to sense and feel? To receive and process requires a line of focused feelings that are being acted upon in ways that control and manage emotion. A breakdown in the sense, feel, and focus cycle limits the ability to respond, because receive and process cycles are disrupted.

Hence, without a line of focus, it is more difficult to receive because the things you feel are not being organized. This is a lack of ability to focus to receive, which results in less information being processed to express a point of view. This is why learning ways to show a learner how to focus is critical. When a learner does not know how to focus their brain, body, and senses to learn, it is because they lack internal control, cannot control their body, and lack sensory organization.

Human Problems

Every problem has an event state that can be associated with a goal state, which will convey any number of possible affects on the self, other people, or the environment. When the brain's body is not set up for contact, then the problem of learning can be found in the behavior that displays the affect. The affect is, hence, the cyclic variable in the exchange of contact between home, school, neighborhood, and workplace situations.

Human problems are problems in learning how to live in home; school, neighborhood, and workplace networks that affect a child's ability and capacity to grow, mature, and develop the necessary human system skills to compete in the process of reflection through the experience of knowledge, and information.

Living in a family that suffers from drug-related problems such as abandonment, neglect, abuse, and prenatal exposure to drugs create deficits that hurt the optimal growth of a child. Hence, a child that has been hurt by major life events in their home, school, neighborhood, and workplace networks may also resist the need to accept help. This forms an interaction problem, where the affect of being hurt may cause the partial or total withdrawal of the sense, feel, and focus cycle. Hence, the act to withdraw is synonymous with the aim to resist.

- For example, resistance is an emotional state in opposition to any action that does not include signs of a defiant behavior challenging life experiences in the face of complex human relationships.
- For example, care is an authentic state in opposition to any process that does not combine personality with character to challenge life experiences in response to complex home, school, neighborhood, and workplace interactions.
- For example, emotion is a complex pattern in reaction to contact that describes the experience of a child in the event as feeling anger, fear, or anxiety due in part, to a lack of information, knowledge, and reflection in learning how to live through these situations.

Progressive Investing Model 50

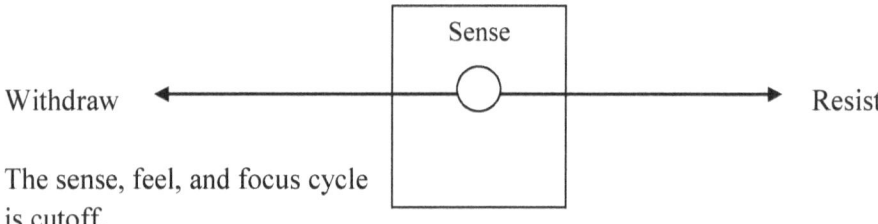

The act to withdraw forms the need to resist, because how the act feels is unknown as to any state of focus or sense making. Hence, when the behavior is fixed, the sense, feel, and focus cycle is cut off. This is because the brain has not been set up to learn through the act to withdraw, which would change the affect or the behavior to an action that would be felt. In other words, learning from the act is produced by the change in state of mind or open path to feel the contact.

Human problems are problems in which contact between the self, the environment, and other people form dysfunctions due to poor school experiences that produce disruptive or disturbed learning, behavior, or conduct relative to a lack of information, knowledge, and experience.

The need to experience school is cut off from the need to learn from the experience of contact at school. This problem deals with being prepared or failing to be prepared to receive the contact between the self, other people, and environment of school.

Learning in school with a poor attitude from feelings of hurt may induce anger, fear, or anxiety from not being prepared for competitive contact in-group situations. Children that are hurt by poor school experiences may hide behind aggressive, intimidating, and bullying personalities where he or she may also seem easy to upset some or all of the time. These are signs of being resistant to contact that require some degree of critical thought.

- For example, if a parent is not prepared for social or academic contact, then the parent's children are less likely to be prepared to learn with a sense of feeling for the needs of self and others in mind.
- For example, the self has to be set up for cognition to make the distinction between the known and the unknown or between being aware and being unaware of how to respond to care.

Progressive Investing Model 51

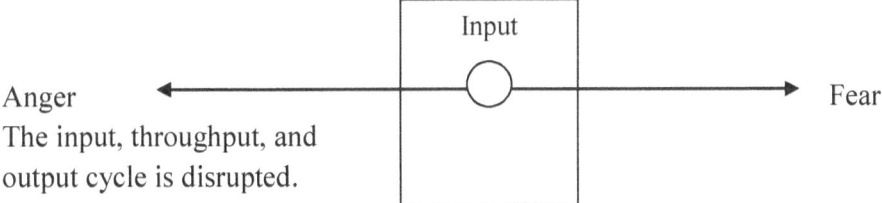

The event of anger is to produce a state of fear as an affect on throughput or the flow of energy as uncontrolled output that has not been felt as input. Hence, when the emotion is not channeled, anxiety moves the anger to produce a state of fear. This means that the body leads to a throughput and output cycle that is resistant to contact or input from the brain, or that the sense and feel cycle has been cutoff. This behavior impedes learning because there is a lack of connection between the event of anger and need to focus how the anger feels. Hence, this would relate the self to the other people and the environment to change the path to one that is guided by the flow of energy to and from the brain.

When there is no positive flow of energy between the act to live and the aim to learn the signs of productivity between home and school, dynamics are affected by the lack of coherence between social enrichment and academic achievement. The act to withdraw produces resistance as an emotion that removes or is at odds with the use of discipline and self-control to secure a line of focus. Hence, there is less of an open path to experience contact.

> ⛭ For example, energy produces the ability to take action and the capacity to learn from the action that is taken, ways to share and exchange new flows of energy to reach complex goals. When contact is met with anger, fear is the goal of the child that has been hurt to reduce the anxiety that is in, not knowing how to respond with care.

Progressive Investing Model 52

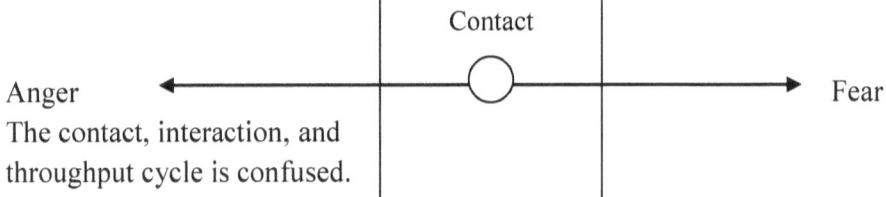

Anger   Fear
The contact, interaction, and throughput cycle is confused.

The anger produces the state of fear through the physics that are displayed in the act to feel the interaction and cooperation with less regard for the contact. In other words, the interaction is supposed to feel bad, which means that the cooperation would look like a state of fear. Hence, the self would be loaded with signs of anxiety brought on by the physical threat of impending discomfort with the way the contact causes other people to feel; meaning, the anger has been transferred. Since fear is the goal state, how the contact feels has been confused by feelings of withdrawal, in that, the self has been changed to a state of anger.

This happens when a person does not know how to use their brain's body to learn or to practice learning from the experience of contact. There is less of a free flow of energy through the human system as a sign of being open to experience. Hence, there are few or less signs of a cooperative state, such as character is a willingness to endure an affect long enough to show gains from the experience.

- For example, learning is a process variable characterized by a person's ability to relate new forms of information to prior experiences through the practice of receiving, processing, and responding to feelings, which produce a pattern of affect.
- Without the connection of reason to the process of learning, the role of logic or a mental effect is more of an affect than a goal state.
- For example, reason is the logical process variable, which leads to the cycle of learned goals that justify the act or choice to participate in the event to live, learn, think, and respond to the feelings in an affect. Focused learning is a logical response to a human system that is not prepared to live in a home, learn in a school, think in a neighborhood, or respond in a workplace.

Progressive Investing Model 53

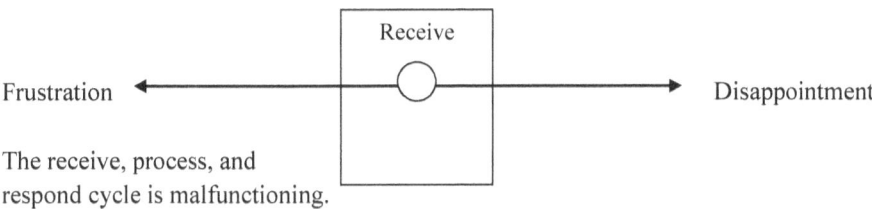

The receive, process, and respond cycle is malfunctioning.

Frustration is a sign of disappointment, which affect receive, process, and respond cycles by blocking the path to receive. When the path to receive is blocked, the sense, feel, and focus cycle is also cut off. Meaning, there is a gap between the act to sense and the goal to focus. If the self is not open to receive, then it is more likely that frustration is a sign of disappointment with other people or the environment of contact. In other words, the signs of disappointment lack a focus, because the brain has not been allowed to participate in the event to receive and process the frustration for its cause.

Human problems are learning problems that relate to breakdowns in how information is gathered, knowledge is organized, and how the self manages to experience thought. Focused learning involves the movement of sense data to the brain for the experience of thought. When there is frustration in the path, there is a denial of the act to practice the transfer of sense data as an open experience, which builds on reflection as another form of thought.

Human problems relate to the breakdown of part or the whole person's human system as a feeling system that focuses on signs of care. Instead of thinking through signs of care, the self resist through a mask of disappointment that causes him or her not to trust. Hence, a hostile behavior is a response to a fear of opening up, to the experience of feelings that lead to thought.

In other words, it is easier to feel bad, because when a person feels bad, they may not have to know how their feelings affect other people in contrast with good feelings that would lead to greater amounts of contact, which in turn, produce lines of focused talk. Hence, the problem of thinking relates back to the problem of how a person learns to live in response to self and others. When a person is hurt, these pathways to learning, thinking, and behaving with care go through a breakdown.

Session 3

Focused Learning Areas of Human Systems Research

Physics	Affects	Mentalism
Focused Learning	Aim	Energy
Progressive Investing	Act	Reason
Sense	Feel	Focus
Discipline	Care	Self-Control
Physical	Emotion	ental
Body	Resistance	Mind
Learn	Learning	Think
Social	Academics	Cognition

Physics

The study of physics is an event for learning from measurable goals how to assess the learner's ability and capacity to focus, which forms a human, cognitive, and behavior science.

Affects

The study of affects is a search to comprehend the effect of goals on signs of feelings to set up process cycles that allow us to assess the experience.

Mentalism

The study of mentalism as an effect is to differentiate human systems science from classical sciences that separate the cognitive aspect of being human from the physical event of being human.

The purpose of the above focused learning areas of human systems research is to expose the actor or learner to the role of physics, affects, and mentalism that make up this school of thought. The participant has to be exposed to the authentic live, learn, think, and respond mode, which forms from the act to experience living, learning, thinking, and responding to contact. Hence, the participant had to be set up to accept the authentic act to receive, process, and respond to sense, feel, and focus contact to interact at higher stages of being cooperative, relative to levels of participating in an organized effort to learn from the study of these events.

Session 4

Stages of Focused Learning

Stage 1: Focused Learning

Focused learning is the physical ability and the mental capacity to aim the body and thus the mind to live through contact. The process of which is analysis, relative to the flow of information as a pattern, practice, or study of human experience.

1. The self is a physical and mental experience.
2. The self has a live body and mind to experience.
3. Information flows through the self as a pattern of, a practice of, a study of, and in the aim of human experience.

Stage 2: Aim

Aim is to point the body to set up the mind with the intent to move through an event to achieve a goal.

1. The self in the body has an internal goal.
2. The self's mind represents the release of internal goals.
3. The self's intent to move is a voluntary goal.
4. The affect on the event is connected to achieving the goal.

Stage 3: Progressive Investing

Progressive Investing is a broad act to live each day to learn how to become more informed through choosing safe areas to study learning from experience in home, school, neighborhood, and workplace relationships.

1. The act to live with the aim is progressive.
2. To learn how to become more informed through the aim is progressive.
3. To choose safe areas, as opposed to unsafe areas to learn from experience, is progressive.
4. To study the act of learning from experience is progressive.

5. To choose ways to study how people live in home, school, neighborhood, and workplace relationships is progressive.

Stage 4: Act

An act is doing something complex to live in a special way to interact with a goal to improve the person or the self, based on reflection.

1. To take action to live is reflective.
2. To interact with a goal in mind is reflective.
3. To improve a person or sense of self is reflective.
4. To do something complex in the environment is reflective.

Stage 5: Sense

To sense is to perceive and process information as a method of becoming aware in relation to tension, pressure, or stress.

1. Recognize and observe that the natural impulse of the body is to learn.
2. The process of information through the eyes, ears, and skin is to learn.
3. A method of becoming is used to learn.
4. To be aware of tension is to learn.
5. When the body moves in relation to stress or pressure, it is to learn.

Stage 6: Feel

To feel is to be physically aware of an experience through feelings of contact.

1. To be physically aware of an experience is to be aroused.
2. The flow of feelings affects states to be aroused.
3. Accepting how contact feels is the process to be aroused.

Stage 7: Focus

To focus is to aim the capacity to channel an affect on the movement of information through the central nervous system.

1. To feel the intent to receive sense data is within the goal to manage physical and mental reactions.
2. To channel the act to control how contact feels is a release of intelligent reactions.
3. The transfer of feelings develops knowledge of the brains possible reactions.

Stage 8: Discipline

Discipline is physical action learning through the use of the body to establish a sense of contact and focus.

1. Social learning is transferred as the skill of the body to focus appears aimed.
2. The body is used in the physical act to learn from the sense of being aimed.
3. To be in control of a focused body establishes that the contact is aimed.
4. The shared physical action and sense of contact means that learning is aimed.

Stage 9: Self-Control

Self-control is the capacity to manage the transfer of emotion through the choice to care about how the body participates.

1. The ability to feel one's sense of self on the outside of the body is connected to the role of focused thought.
2. The transfer of emotion through awareness is a technique to improve the role of focused thought.
3. The choice to study becoming aware is to reinforce the role of focused thought.

Step 10: Physical

Physical is the social characteristics of the body and the senses, but through the focused use of energy, space, time, and agility.

1. The human features of the body are social in relation to reality.
2. The senses are the tools that connect specific functions to reality.
3. Signs of strength, matter, age, motion, and skill act as fields of reality.

Stage 11: Mental

Mental is the invisible processes of the brain in relation to acts to sense, feel, and focus signs of self-control.

1. The subjective events of how the brain works are known through receive, process, and respond cycles to live, learn, and think.
2. The act to experience contact is known through receive, process, and respond cycles to live, learn, and think.
3. The connection between sensing, feeling, and focusing is known through receive, process, and respond cycles to live, learn, and think.

Stage 12: Body

The body is the physical human system of a person.

1. The human system represents the social structure of a human being.
2. The human system reflects the whole human being.
3. The human system relates to a framework for being identified as a human being.

Stage 13: Mind

The mind is the place in the body for mental connections to a person's human system.

1. The mind is the place where intellectual functions lead to learning.
2. The mental connection is the psychic process that leads to learning.
3. The self is organized to use the human system, which leads to learning.

Stage 14: Learn

To learn is the act to try through the study of external experiences, such as reading, listening, smelling, talking, and feeling.

1. The act to try forms the experience of an affect.
2. The effort to study the experience forms the experience of an affect.
3. To read, listen, smell, talk, and feel as relative functions form the experience of an affect.

Stage 15: Think

To think is in the aim to care about the act in which to interact is an experience in the role to cooperate as the path to participate with care.

1. To care to communicate is a clear sign of reason.
2. When the act to interact appears aimed, it is a clear sign of reason.
3. To be able to handle the act to focus is a clear sign of reason.
4. When a cooperative participant acts under the influence and formation of the process in thought, it is a clear sign of reason.

Stage 16: Social Enrichment

Social enrichment is the physical experience of care delivered through interaction with other people in the exchange of help.

1. The enrichment is the physical aim to care through the act to improve contact to influence a sender or receiver relationship.
2. The experience is modeled through the process to interact with other people to improve the influence of a sender or receiver relationship.

3. The exchange of learned behavior is specific to the aims and roles of a sender or receiver relationship.

Stage 17: Academic Achievement

Academic achievement is an outcome of formal learning characterized by special signs of discipline and self-control as states of responding to instruction through the act to learn as a successful learner, student, or researcher.

1. The description of special stages of accomplishment is represented by the ability to learn how to live in a home, learn in a school, think in a neighborhood, and respond in a workplace.
2. The signs of discipline and self-control are specific to a set of skills such as to appear aimed and focused to learn how to live in a home, learn in a school, think in a neighborhood, and respond in a workplace.
3. The formal learning of character is a physical and mental act that distinguishes the person's capacity to learn how to live in a home, learn in a school, think in a neighborhood, and respond in a workplace.

Session 5

Reflection: Focused Learning

Focused Learning is a Human Systems Research Program designed to improve the lives of children, youths, and young adults that have been hurt by major life events in their home, school, neighborhood, and workplace networks. Focused learning theorists and Human Systems Researcher Dr. Christopher K. Slaton believes that our public schools need to teach to a child, youth, or young person's brain, in contrast to the body, to prepare him or her to participate in the process of emotion.

Focused learning advances the goal of human awareness through the disciplined study of how children that have been hurt live, learn, think, and respond to contact. It also advances the aim and role of human science and builds on science-based practices in which helping professionals can study, look at, and think about new ways to recover and support high-need groups.

The help the helper help system offers a new research community to solve human and personal problems and to be more informed as human systems researchers. When a school of thought chooses to participate in a program to improve how humans learn as a system, they are more likely to come to think about the whole potential of the self as a human system.

Reflective Progressive Investing Model 54

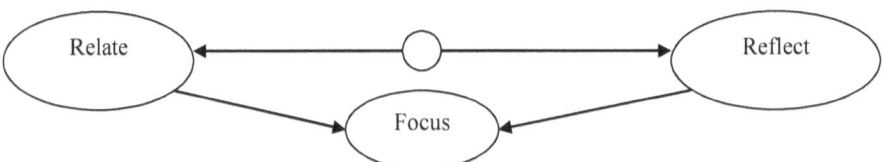

To relate is the event, which connects to the process or goal to reflect that has an affect on focus. Focus, as a growing state of mind, flows from one's aim to relate to the experience of contact. Hence, this is an act to reflect, which has an affect of how the self comes to a state of focus through the process of the aim.

We have learned to relate the act to live as an event that lead to stages of focus, as an affect of the mental goal to reflect, which recycles the act to live to learn how to use the brain's body.

1. To do this, we learned a new format to solve human problems in the twenty-first century.
2. To do this, we learned a new approach to working with children known to be hurt because of having been born into a family that abuses drugs.
3. To do this, we learned a new approach to teaching children how to live through, learn through, think through, and respond through a higher sense of self as a human system.
4. To do this, we learned a new way of thinking about the child, youth, or young adult as a human system rather than as parts to a whole.
5. To do this, we learned to set up more leaders for the role to focus: observe, learn, listen, help, and lead through open experiences of behavior, personality, character, and attitude.

Workshop Notes:

Now that you are experiencing focused learning, tell us how you feel.

Tell us what is in the experience of focused learning.

Select one (coach, teacher, mentor, counselor, and participant) role and tell us how you experience the effect of being a participator.

Submit your request for additional sessions.

☐ I would like to set up a date, time, and location where my team and I can receive more Progressive Investment Training.

My name is _____
My title is _____
My e-mail is _____
My mailing address is _____
City _____
State _____ ZIP _____
My preferred dates are _____
My preferred times are _____
My preferred location is _____

☐ On-site
☐ Off-site

FOR A BUSINESS REPLY

Use the self-addressed envelope that is enclosed with this guide.
Mail envelope to the following:

Progressive Investment Group
Progressive Investment Training
ATTN: Progressive Investing Institute of Focused Learning
P.O. Box 278363
Sacramento CA95827-8363
slaton@softcom.net

Epilogue

Systems Feeling Theory

How do you feel things, and do you feel things? Systems feeling is a new science on how the brain, body, and senses feel things. If you want to learn why the body, senses, and brain are wired to feel things, you have help. In this final section of Education and Science, the Information Processing Age: the Learning Parent and Child in Crisis, you learn to move with your feelings. You can learn to deal with how you feel, because of the way energy, action, and feelings move the body, brain, and senses to live, learn, think, and respond.
You have to choose the right path.

Teaching to the Brain

In most acts to live, there is a process cycle that expresses how the act feels through a set of components (Kaufman 1980) that relate to the movement of energy as interactions form a complex flow of action we can observe as feelings capable of being defined as a system. Hence, human systems feeling refers to the use of the senses to feel, focus, and send data to the brain to be received through process cycles in the body. The study of human beings as systems, then, is viewed as science-based learning, in that we know very little about how the brain, body, and senses form as a set of components that serve a common aim—to live, learn, think, and respond to self, other people, and the environment as a force.

This is the process we describe through human systems research as forming practical ways to study self action that relates to the acts of other people and their actions in the environments they all share as a man-made system—homes, schools, neighborhoods, and workplaces. Energy forms as action—which, in human beings, form feelings as patterns of expression described through the use of language: being cognitive, physical, mental, social, environmental, emotional, sensory—and a technology that lives through the use of science. Human systems science seeks to find the meaning of order in the basic processes that relate these words to the human experience of uncontrolled and controlled energy, action, and feelings through human systems thinking, or from the way humans feel things to think from the inside to the outside world.

Now, we can make better sense of our systems theory (Kuhn 1975) in relation to human behavior as the release of energy, action, and feelings through relative, interactive, social, and perceivable patterns that express the experience of contact (Anderson and Carter 1992). We think, then, through the use of a mind that becomes more mental as stages in process cycles that take in input to sense, receive, choose, and send forms of thought from these feelings. This is called a feeling system that influences the production of thought in the brain (Amen 2010). We express this as the appropriate way to teach children how to learn from self, other people, and the environment (Slaton 2002, 2009). Children are "feeling systems" that learn to grow from the experience of contact based on the things they are able to do with their brain.

How we educate children through the use of science to process information needs to be connected to the brain, body, and senses as a human system. This is a method in the age of information processing to improve learning and support services from the basic act to live. This would help teachers learn how to apply a human systems approach to the work they do with children, youths, young adults, and parents in crisis. This is a complex formula to address the difficult subject of how to teach to a child's brain.

The first step is to set up a process, to screen and assess young children for behavior, learning, and mental health concerns at the preschool and K-3 levels. We will be able to learn how the behavior relates to learning and why they both affect mental health.

1. Behavior is related to issues in the home.
2. Learning is related to problems in the school.
3. Mental health is related to concerns about neighborhood, school, and home interaction.

Second, brain, body, and sense functions are set up by inner and outside connections. Human systems science is a logical way to gather information and to set up a body of knowledge from the experience of thought through reflective practices. Hence, the act to live is analytical. At each level, where contact and interaction occurs within the system of systems below, each set can be analyzed as a way to advance how teachers teach to a child's brain.

Study the act to live: Level 1 Environmental Learning Systems

1. Input Energy
2. Throughput Action
3. Output Feelings

Study the act to live: Level 2 Social Learning Systems

1. Sense /Receive
2. Feel/Organize
3. Focus/Express

Study the act to live: Level 3 Academic Learning Systems

1. Receive/Detect
2. Process/Select
3. Respond/Effect

Study the act to live: Level 4 Human Learning Systems

1. Choose/Accept
2. Comprehend/Learn
3. Change/Participation

Study the act to live: Level 5 Green Science

1. Send/Information
2. Receive/Knowledge
3. Reprocess/Experience
4. Feedback/Reflection

With a screening and assessment process, behavior issues can be caught before they become learning problems that affect a child's mental health. When you think of the child who is born to a mother that abused drugs while they were in utero, each day, they have to act to live in ways that will not trigger negative events, goals, and affects. A child that has been hurt by major life events in their home, school, and neighborhood has to learn how to do the things they can to become more informed from where they are in the problem situation (Slaton 2009). Each year, more and more children are moved through the public school system lacking the learning and support services they need to adequately compete.

The screening and assessment process noted above is a science-based learning tool kit that allows a parent, teacher, or other helper to test the child's send and receive paths. When you are a family leader, you have to be an educator also. We care about how a family lives and learns how to use education. When you live in poor home, school, neighborhood, and workplace networks, you need to know how to lead and learn to move through these places with signs of care. Hence, you want to live each day to learn how to make your life feel better. You can help save a child's life, by talking to their brain.

The human system is set up to feel things as an open system (Bounding 1956) to live through the experience of contact. We learn how a child in crisis may function as a consequence of being affected by a negative process variable that alters the flow of feelings, which lessons the degree of control. Hence, a child that is growing up hurt will seek to find order through conflict to resist a state of calm for a state of chaos and less acts of control. This means that we think teachers need to use a systems model with a Progressive Investing approach.

The Nature of a Systems Feeling Theory

Space is never empty, time is never stopped, and energy is never held back by the same rate of care. This means, the qualities that make up the spirit of the earth are global. Life is such that we can find parents and children all over the world out of touch with each other as though unable to make sense of each other's human act. Space can never be empty, because there is tension all over the world from this disconnect, from meaning. Time never stops, because the pain of growing up hurt keeps spreading stress all over the world. Energy is never held back by the same rate of care, because we feel this pressure all over the world.

Where has our sense of environmental awareness gone? The stages of our initial contact as environments set up for input, throughput, and output through the flow of environmental experiencing has been upset by the lack of a learning state. This is at the base of the human system, where energy moves action through feelings to help us learn how to live as level 1 learning. The experiencing of the environment is intended to create awareness for the needs of the environment of self and others, which links us to what it means to be human and to be a living organism. We learn this through the experiencing of the experience of our contact, which results in experiencing this effect.

The senses are armed by our acceptance of the initial contact, which move us to a stage of continuous interaction. Sense perception is the social search for cause and the meaning of the way it feels to have feelings moving through the human system from our initial contact with time, space, and energy. Hence, emerges the natural relationship between our act to sense, feel, and focus brain and body events as level 2 learning. To do this, we allow the energy, action, and feelings to move us through the uncontrolled state to an internal need for a controlled state, in which the path of focus guides the continuum of interactions through perception. The mind and body experience emotion to become more aware.

This is because the cognitive state is created at the initial experience of contact and the sense, feel, and focus cycle organizes the mental process for level 3 learning. This is the stage where cooperative processing is needed to bridge the environmental learning system with the human learning system to receive, process, and respond to nature and man-made demands on brain, body, and sense events. Recall that both the earth and we, as human beings, are natural events to experience, but that anything made or created by human beings for other human beings to experience is a mechanical event. For instance, academics are not natural noises, sounds, signs, and symbols for the brain, body, and senses to experience. Hence, many children experience a negative reaction to the energy, action, and feelings they are intended to cause.

The purpose of physical action learning is to help set the stage for cooperative processing by formatting the brain for the experience of this potential tension between natural science and man-made science. Hence, this is the stage of participative systems, where the physics of being human; signs of care are related to the action learning elements. Recall that the senses are physical and that the act to receive sense data is abstract functions. The base system at level 1 learning move us through the events of life and send the free flow of energy, action, and feelings as input, throughput, and output from the uncontrolled environment.

This means, what we are and what we can become are at stake in level 3 learning.

To participate in the continuous acts of learning how to live with self, other people, and the environment, what it means to be able to live, learn, think, and respond has to be realized to choose willingly and freely to comprehend the need to change and accept change as a learning cycle. This is level 4 learning. The fact that humans have the ability to make choices was left out of the behaviorist's theory on human behavior. You can remove the option to choose in the initial experience of contact, but you cannot stop the human brain from learning to adapt to man-made consequences. As the brain interacts with the body, the body loses its uncontrolled energy, action, and feelings. Hence, the brain shows signs of care for the body through a willingness to comprehend the need to change.

What we have learned is that a child will become whatever they must to live with us or without us. The difference between a child and an adult is that the child may not know they have the capacity and potential to choose the need to comprehend the things they can to accept change as a state of becoming. However different they may seem, some adults do share this predicament of not knowing how to experience changes through contact and interaction. Then there are those in both groups who do know how to move through uncontrolled and controlled states of energy, action, and feelings that produce tension, stress, and pressure.

These are both, parents and children, and children and parents that have learned how to reprocess the way it feels to move from the known to the unknown, with less fear of being found not to know, what they are suppose to know, how to do with academics. This is the stage of emergent states of becoming a green science in level 5 learning. Progressive Investing and human systems science show these stages of emergent states through interdependence and a multidisciplinary systems approach. This helps us

- identify events that a system of interrelatedness must be set up to experience within separate but connected boundaries;
- analyze and define the possible ends of each network to understand the interrelatedness to each element of the whole system, particularly the possible flow of feedback in cause, effect, relationships (how something that was once an affect transcends to become an effect, and ultimately the behavior); and

- synthesize to classify the conditions that produce or affect the flow of energy, action, and feelings over space, time, and new events

This is a human systems research framework relative to learning how to live as a Progressive Investing goal:

- Determine the points at which choices can be understood as being available to the human being as they make contact and interact at the various levels of the systems and the inherent affects or trade-offs that result from the event or choice(s).
- Identify short and long-term effects in the affects or trade-offs within the action learning continuums of the system(s).
- Decisions are based on interpretive research findings and action is taken through an understanding of the participant's role or function and goals in relation to the affects or trade-offs and the accumulation of information, knowledge, experience, and reflection within time, space, and events used to evaluate the whole system.

These are the aspects of human systems that have been recovered from crisis and are now in a future state of reprocessed cycles of experience through reflective practices. When the sense and receive paths move successful flows of energy, action, and feelings to the brain, it increases order. Each time the brain works in the reverse of the body to pull energies forwarded, it is to meet the internal need for the brain and body to work through order and structure. Hence, all forms of the self—behavior, personality, attitude, and character—are drawn together as a working whole. The experience of Progressive Investing: a practice to take action and to become more informed over time forms through the experience of contact and interaction and sets up this social, psychological, emotional, and physical effect.

The patterns of wholeness that become more known as the self becomes more aimed in the relationship of learning how to live with other people's fuse and order becomes based in the structure of these effects. These are the dimensions that connect new understandings of human, cognitive, and behavior sciences to real life acts. The points we have designated—the event state, the goal state, and the affect state—refer to this method of discovery, the received to the unfamiliar system of coordinates. The experience to live each day to become more informed moves this vision, care, and aim through the act and practice to learn how to be a leader and a follower of thought.

The Learning Parent and Child in Crisis

As a parent, you need to become more informed about how to educate your child for the global economy. In the global economy, race matters, but race cannot be the thing that breaks down your child's ability to sense and receive data as free of emotion as is possible. This means you need to live with a plan to generate sense and receive skills that allow your child's human system to function as a technology. As you look back on these words, *green science*, they mean that your child is an emerging twenty-first century technology. Hence, you teach a child how to use the environment to live, learn, think, and respond to self, other people, and their environments to show them how energy is reprocessed into higher states of becoming more informed.

The transfer of information, knowledge, experience, and reflection on how to move through people to improve the way you use a home to lead, a school to help, a neighborhood to listen, a workplace to learn, and the human system to observe gives you and your child the skills and strength to compete. This is the Information-Processing Age, where the ability to process knowledge is critical to the experience of how you learn to become more informed through the way it feels to look back and be able to focus the things you can.

This is how you learn in a time of crisis. Growing up hurt has its own behavior. The good learner gets to go to Penn State, and the poor learner gets to go to the state pen. It is the words that need to make sense to the child to strip away the nonsense about the affect of being hurt. A child's behavior at home and at school has to reflect how aware they are in social settings about ways to live and learn to act through crisis. Deviance of any kind breaks down their aim and control.

All you need to do to rescue a child that has been hurt by major life events in their home, school, neighborhood, and workplace network is to learn how to live with them, learn with them, think with them, and respond to them. Drawing from more than twenty-five years of experience with children and families in crisis, Dr. Slaton uses evidence-based practices to show you how to set up a human network. You need to find out how to assess your child's school in a fair and objective way, and on the basis of how school experiences affect your home.

Your child's brain is wired for learning how to live, learn, think, and respond as they grow. Learn how to use these words to improve your child's ability to mature into a learner at home, at school, and in their neighborhood by talking to their brain. Learn how to set a child's mind up to live in your home, to learn in school, to think in a neighborhood, and to respond in a workplace. We prepare parents to practice doing the things they need to do to develop a child's capacity and potential to sense, feel, and focus their brain with care.

Every parent has to learn how to move a child through school. Children who struggle with reading, writing, math, and behavior skills need to learn how to act through the problem. Learning how to sense, feel, and focus their brain to receive, process, and respond before, during, or after they have been hurt is a human systems approach to reset and correct the way they experience self, other people, and their environments. Learn why not being prepared for contact is a key factor in poor performance and find out how to structure a home, school, neighborhood, and workplace network for a child that has been hurt.

A child needs to learn how to learn. This takes us to four places parents are taught how to use to measure the quality of their child's effort, focus on positive values, openness toward academic goals, and self-respect over time. Parents get in-depth knowledge on how the brain, body, and senses influence language and cognition and leads to physical identity and the mental process. Progressive Investing models chart, organize, and provide easy-to-follow steps on ways to adapt how you live, learn, think, and respond to feelings. Hence, every child can learn how to learn to use their brain at home, at school, and in their neighborhood using the system, strategy, and process we teach to parents.

Learn how to set up your child's brain through their natural acts to live, learn, think, and respond and reduce the time and stress of not knowing how to feel things as events, goals, and affects. A thirty-six-hour course guides parents through the learning process to acquire a Community-Based Learning Certificate to deepen an understanding of content, word use, and from a real life analysis of what matters. We guide you through proven, evidence-based practices to share the skill sets you need to decide how to help your child learn through the crisis of growing up hurt by using these interventions in your home.

Make the home the start point in your fight for a better way to parent, and the idea alone will change how you use it to live. Behind this perspective is a way to help parents learn how to become better leaders to achieve positive changes. Learn how controversial parents as leaders and child developers have to be to nurture and move their ideas as advisors and advocates to coordinate, collaborate, and build a small learning community of helpers. These strategies start in the home as a safe place to learn who is interested in helping children that have been hurt by major life events in their home, school, neighborhood, and workplace networks, and in human networks.

Because your child loves you, all you need to do is learn how to thrive as a parent leader and child developer. Every parent who learns how to lead a child learns how to lead a family, learns how lead close relatives, and learns how to mentor other parents. Learning how to be a parent leader and child developer is not a program. It is a process you have to learn how to practice each day. How you become informed, how you become more knowledgeable, how you become one with experience, and how you become more scientific through reflection are basic ways of learning to lead. This school of thought offers formal and informal ways that teach parents with children at every grade level to expect results.

When you know how to assemble the right team of helpers, develop a successful process to live, learn, think, and respond through Building Human Asset Meetings to reduce conflicts, to motivate teachers, and to help build more persuasion skills for the team to experience. During simple weekly meetings in your home, talk about reading, writing, math, behavior, and problem solving from the social perspective to make it easy to sense, feel, and focus academic concerns. Instead of feeling disconnected from your child's academic issues, we offer ways to use reality-based learning tools, games, and role plays to explain why children who grow up hurt need special learning and support to help them adjust to meet these odds.

Once you get started, we show you how to adjust and tailor life needs so that each person can explore ways to make contact, interact, cooperate, and participate through the use of talk, memory, self-awareness, and thought to structure the way they learn alone and with the team. The home becomes a richer place for your child to learn about you, with you, and from you through learning events that create richer signs of care in a context where feeling things carry a lifetime of effects. Reality-based learning keeps your skill sets up-to-date, because so much is shared from the way a person enters your home, sits down, makes eye contact, uses talk, and looks for signs of care. You learn from a twenty-first-century curricular and instructional approach that brings out your needs, because you are being treated with great amounts of care by human learning consultants.

Explore neighborhood learning strategies that get parents and children involved in learning how to improve life inside the home and foster success outside the home. Use essay writing to host reading and writing contests to connect your child to the spoken word that results from learning how to write to the brain. Learn how to connect what you do in the home to the things your child has to be able to do in school through art contests to help him or her get a feel for acts to sense and receive data. These techniques are good for both you the parent and the child to move through learning how to live, learn, think, and respond together over time. Any parent who wants to stop acts of bullying and teasing in their neighborhood will find ways to use song and dance routines to help children learn how to clean up their act.

In this approach, parents learn how to act as helpers, advisors, advocates, and mentors because they need to reduce signs of hurt before they lose their child's focus. Using this plan is not easy and does require focused learning skills to explain what you feel, how you feel, and why you feel the way you feel to feel your need to know how to lead and develop your child's right to a fair and reasonable education. Learn how to apply research-based practices to improve home-to-school learning in ways that make it easier to screen and assess how you live, learn, think, and respond to self, other people, and their environments. These are the key words that influence the actions of people though the way you use talk and aim this approach.

Each year, your work will be supported using new literature and research-based tools, events, and goals, which you can use to improve your home and family as a business. You learn to choose meaningful roles for your child to act through, because you know more about their academic strengths and weaknesses to improve how they function alone, in groups, and in competitive play events. The practices that work in your home will translate into action with a clear blueprint for success outside your home. Human Systems Research is a natural and authentic way to learn from the study of self, other people, and their environments ways to inform, instruct, govern, and lead a family.

Based on the concept of living each day to become more informed, you learn the steps parents need to take to produce action plans that connect the four most important places a child must learn how to move through as they grow, mature, and develop. You get real proof that measures how well your child is moving through the crisis of growing up hurt from the use of parenting strategies designed to help the helper help. To learn how to help is a perfect step for parents who rely on self-help or family and friends to make it easier to practice parenting from an applied perspective. What moves you as a parent moves through them as helpers from a Progressive Investing perspective that works. Hence, your work is aimed to maximize the benefits that flow from everyone's act to live through the crisis to help.

You have been reading a book that spells out ways to learn how to live with a child that is growing up hurt by major life events that affect their ability to sense and receive contact. To learn how to learn how to lead while learning ways to overcome deficits that hurt a child's ability to perform free of a poor behavior is why you have reached this point in our talk. You know how important parents are in each step a child has to take to achieve success in school. But what this school of thought does with parents is lead them to improve their direct influence on children as they act through parenting roles to compete. With a growing knowledge base on how a child that has been hurt learns to move through a poor behavior, your work as a parent will be the key factor that reduces their anger, fear, and anxiety.

Living Each Day to Become More Informed

The best way to summarize this work is through our talk to a parent in crisis that called us for help on her child's first day of school, August 16, 2010. She had been notified by the school that her child was missing and that the police had been called in to help locate him. The child had gone to the wrong class. The issue for the parent, though, was the call and the mobilization of her family network. Family members left their places of work to come search for the child. The police officer assigned to the parent's home said the teacher that allowed the child to sit in the wrong class ought to be fired.

The parents concern was for the behavior that the child had regressed back to over the summer. She allowed him to go stay with relatives for only two weeks. After two days, he was constantly calling her. Within the first week, he began to ignore rules and disrespect his uncle and aunt. By the time he returned home, he had reverted back to throwing tantrums, acting out in public places, and refusing to control his anger. Parenting a child that has been hurt is a skill.

When a child is hurt by major life events in their home, school, neighborhood, or workplace network, you have to talk to his or her brain.

You want to help him or her move toward a calmer state of mind with less openness to chaos. This means, you work through the chaotic state by providing a structured approach to the request you make in relation to the behavior and the changes that must be made. This is called teaching your child how to live and learn with you in the lead. In this process, you make contact to interact through dual states of cooperation as a set up to help you provide a structured approach to helping him or her help you improve their capacity and potential to participate.

You work through the tantrums, acting out, or signs of being upset with the use of special words and talk to his or her brain. You have to learn how to aim the words and to choose the words that allow you to appeal to your child's real sense of self and the world. This is the bridge between acting out because of uncontrolled emotions and acting out because of underlying mental health factors. In either case, you want to bring your child's sense of self and the world into a more organized feel for discipline and self-control through the use of special words. You measure responses in relation to your contact, as you talk to the brain, and then measure responses to self and other people.

This means that you are working to move your child through the act to interact, which is where the process variables of uncontrolled emotion and conduct may be crossed up in a resistant state of hurt or anger. You have to take a serious look at your child's body in relation to your child's brain and the way the senses are used to receive data to organize and express feelings. You learn how to feel for your child's state of mind to help move them to learn how control feels through your effort. You provide meaningful examples for your child to relate with as you challenge him or her to use their brain to control how they feel you.

By setting up meaningful paths to talk through, signs of care can be aimed as a parent's role that coincides with the need to lead the child to receive help to work through the way they feel you. When you are a single mom, your work with a male child that is hurt has to be set up to move them to feel things in the sense of how you want them to come to think with real people in mind. Your voice, experiencing you, and the way you look become mental products and, hence, are social acts in the broader context of parent-and-child interaction. This requires learning to set an environment up to be experienced, or for the experience of your child, in terms of your values, beliefs, and norms.

The way you use talk to set up events, the goals of these events, and the affects you use to measure possible changes in behavior or conduct have to be shared. This is a parent's list of suggestions and/or recommendations on how to communicate with a child that is growing up hurt. You live each day to learn how to inform them.

1. Live with me.
2. Learn with me.
3. Think with me.
4. Respond with me.

You talk to the child's brain. Today, more than ever before, young people need structure. When a young person is growing up hurt, they learn to live with you through the structure of these events, goals, and affects for them to experience. They live through the state of calm in terms of a willingness to work within certain rules of behavior that allow you to work with them to reduce or eliminate signs of anger, fear, or anxiety.

This will change the way parents parent children that are growing up hurt. These events emphasize the process of living to learn how to work with the child's brain to identify problems. You do this work to rule out myths about whether or not the child can learn through the ongoing study of his or her capacity and potential to move through contact. Hence, the parent comes to recognize the importance of sense and receive paths being open and free of a resistant state. Once the effect of talking to the child's brain has taken root, his or her response to feeling things through should lead to the experience of a calmer sense of self and the world.

1. Can you live with me?
2. Can you learn with me?
3. Can you think with me?
4. Can you respond with me?

When you apply the word *home* to the word *live,* the act to live connects the process of the home as a social place. The place a child will need to feel these words is in the context of home and learning how to live. When learning how to learn to feel sense data, your body, your energy, action, and feelings need to be received or felt a certain way. When a child's brain is hurt, you have to aim the things you say. You have to say the things you mean through words that will move them to care about how they feel things to live, learn, think, and respond. Each day, the choice is yours. How you approach each day is up to you.

Parent Education that Works

Learn how to use research-based practices, because no one expects you to be interested in going back to school to learn how to parent a child with behavior issues. You take this step to learn how to help your child, and it deepens your understanding of a parent's role and influence over a child in crisis. This is an ideal way for parents to move through the pain and hurt of living with a child that is less responsive, affectionate, open, and aware of the things they do to help keep them safe as they move through childhood. For this reason, the workshops are rigorous, because they are reality-based. You learn how to feel the pain, talk through the pain, move through the bad memories, search for a higher sense of self, and be more thoughtful and focused. After all, parents need to have a plan to survive the crisis as well.

In this case, your Community-Based Learning Certificate of Achievement is what matters most. The synthesis of the steps you took to move through the crisis and get your child to graduate from high school with a diploma, matter to this end. The conditions and the states of mind you had to move through are real and proven to be the most crucial steps in working with a child that is hurt to increase their signs of care. You will recall how in the start of your journey all the conditions of being hurt caused you to isolate to protect your state of mind—such as crying out through streams of sadness, shame, and disappointment—into a caring and compassionate response to feel for your child's frame of mind.

These parenting classes prepare you to accept the things you cannot change, but to work on the things you can to enhance parent achievement as a framework for home improvement. Your effort unfolds as an action to change how you feel things to plan the things you do to move through them and to improve learning and support services to your child. Your parenting classes relate to your family-centered learning practice, which is a step-by-step approach to why living to learn how to help a child that is hurt moves you through the experience. You learn from the experience of experience, which shows you how to sense and receive the process. Understanding the process to becoming more informed, knowledgeable, experienced, and reflective helps you solve the problems you can and inspires your search for ways to move through how you feel things more prepared to address your child's needs.

Glossary of Terms

Academic Problem. This is not due to a mental disorder or, if due to a mental disorder, is sufficiently severe to warrant independent clinical attention. When there is a pattern of failing grades of significant underachievement in a person with adequate intellectual capacity in the absence of a learning or communication disorder or any other mental disorder that would account for the problem (DSM-IV-TR 2002).

Academics. This is the mechanics of learning formal skills through the use of school to develop literacy in reading, writing, and math. When a person is able to communicate through brain, body, and sense coordination, it is because they are set-up to receive orders. This is a method of learning to use books to read, write, speak, and count. To become educated through the use of school learning is an intellectual skill.

Academic Skills Disorder. This is the problem when learning to receive noises, sounds, signs, and symbols are not processed into meaningful speech, words, and numbers because of difficulty in the sense, feel, and focus cycle in response to contact. When a person is not able to translate contact into interaction that moves the meaning of information through the use of language, they are less able to perform academic task.

Act. The affected variable between the event to look and the goal to think as described by Stringer (1999) as a "look, think, and act routine (p. 19). To act is a role, to look is a function, and to think is the process variable which determines the effect of the practice. When the body is observed as an event, the act to live sets up the brain as the goal to focus, and to think is affected by the act to learn.

Acting. What distinguishes human beings from other living things through the ways we can act (Szasz 1990). Acting takes place in the dual sense, in that we can act to behave and we can act to perform (p. 222).

Action. Energy and changes in signs of motion. In humans, action creates a pattern of energy through the acts to live.

Action Research. This is a science and practice designed to deal with social problems, while learning to develop changes in behavior. (Bentz and Shapiro 1998). This relates to the use of research to improve human action and participation in society.

Adulthood. This relates to the transition between being a minor and the age of 21; parentage, the point in the life cycle that involves the ability to reproduce and responsive action.

Affect. This is the interactive process variable between the act to live and the act to learn, which expresses how a person feels as they move through an event toward a goal. When a person feels contact, how it feels influences external and visible feelings. Some examples are: pain, pleasure, anger, joy, anxiety, excitement and son.

Attachment Disorder. This is a condition usually developed during infancy and early childhood which corresponds to a pattern of social and emotional anxiety that affects signs of normal bonding. When a person experiences trauma at any stage in the conception of a child; the embryonic stage can be upset and the child may show signs of fear and anger when separated from specific caregivers.

Awareness. The process between the experience and a reflective capacity that leads to a higher sense of self, other people, and real things in the world.

Backward Feed. When sense data moves from contact with the body to the brain.

Behavior. This is a physical pattern of response to tension, stress, and pressure which forms the base that shapes character, attitude, and personality as a behavior to contact. When a person is anchored by a healthy sense of self, the body is a sign of an act to move in relation to the needs of the environment as the process to live, learn, think, and respond unfolds.

Body. This is the physical structure of the human system. When a person is born, they are a physical, mental, and social experience to self and others because of the bodily aspects in being human.

Brain. This is the central nervous system located in the skull. When a person receives contact, the senses connect to nerves that transfer how it feels to the brain.

Building Human Asset Meeting. This is the process used to gather and share information. When a person receives learning and support services from Save Our Youth, they become participants in the Building Human Asset Movement to improve learning and support services to poor, minority, high-need and special-need children, youths, young adults, and parents.

Building Human Assets. This is the vision within the process of working with children, youths, and young adults that have been hurt by major life events in their home, school, neighborhood, and workplace networks; the organization of the senses will lead to the organization of feelings, which will in turn lead to the organization of focus and action that allows the person to live through issues in the home, problems in the school, and concerns in the neighborhood more effectively.

Business. This term is related to family, education, and government affairs through a systems theory developed by Dr. Dolores Slaton. When a person has a sense of self, other people (family), and the environment (home, school, neighborhood, and workplace), they are more prepared to interact and compete.

Care. A sign of strong feelings of thought for self, other people, or things in the world.

Change. A variable of some object, thing, or issue that reveals how energy as action transforms as a process cycle in a state of becoming. To adapt, as in not to stay the same, to appear different over time, to be flexible and unpredictable are all signs of how a child may grow, mature, and develop as a person.

Character. A mental, physical, and social display of self in a highly regarded way. In human systems research, character is the stable signs of an ability to lead self and others, through one's behavior, personality, and attitude.

Child Developer. A building human asset term used to identity with a parent's role to care for how a child grows, matures, and develops under his or her charge. In the practice of Progressive Investing, a child is set up to become a learner and participant through the parent's leadership and development.

Childhood. This relates to the opening stages of life between being born an infant and age 12. In relation to parentage, this is the point in the life cycle where the child is most open to learning how to live with self, other people and their environments as a human learning system.

Children and Families in Crisis. This is the cause behind the effort to improve learning and support services. When a person is not able to live in a home, learn in a school, think in a neighborhood, or respond in a workplace because of poor contact, interaction, and cooperation skills they are in crisis.

Community-Based Learning. The program variable in the Progressive Investing approach to living each day to learn how to become more informed through the use of home, school, neighborhood, and workplace networks. This has become the base for human systems research on how humans live, act, focus, and work.

Community-Based Learning Academies. This is the purpose of human systems research. When a person is hurt by issues in the home, failing grades at school, negative influences in their neighborhood, and a lack of occupational skill sets for the workplace; they need a learning academy that works with him or her from where they are in the issues, problems, and concerns that threaten future economic opportunities.

Community-Based Learning Certificate of Achievement. An award offered to parents, health care providers, educators, mental health professionals, social workers, and corrections professionals who complete the Parent Education and Resource Coordination Services Workshops. Professionals who work with children and parents in crisis must possess a high level of expertise in the areas of human systems science problem solving methods, human systems research, assessment in learning and support services, Progressive Investing models, and interaction techniques.

Conduct. This is a pattern of responding to contact in which a frequent state shows a lack of internal control through aggressive, careless, deceitful, and defiant behaviors. When a person is not able to show signs of discipline and self-control that correspond to self, other people, and environments of contact; they show signs of frustration (anger), disappointment (fear), and hostility (anxiety); that relate to breakdowns in their home, school, neighborhood, or workplace networks.

Contact. This is the process variable in learning how to live, used to study the connection between self, other people, and the environment. When a person is born, they physically interact with through acts to live, learn, think, and respond to how they feel; to being touched; and to being exposed; and to issues, problems, and concerns.

Cooperation. This is a goal state in the act and process to learn how to live. When a person is engaged in contact that leads them to interact, the action behind their efforts is to learn how to live through it. Hence, the processes in their capacity to sustain it may only occur through self-help in relation to the receipt of help from other sources.

Crisis. A major life event that threatens to disturb, ruin, or hurt the way a person lives from one day to the next through control over his or her affairs or.

Crisis Intervention. This is a strategic action plan to improve acts to live, learn, think, and respond to tension, stress, and pressure. When a person is set up as a human system, they are involved in learning how to participate from the inside, to the outside of the body to manage and control the release of feelings.

Delinquency. This is the term used to describe learning, behavior, and conduct problems, issues, and concerns that lead to patterns of defiance at home, poor grades and tests scores at school, and deviance in the neighborhood. When a minor chooses not to live with rules, not to learn with other people, and not to think with care for self and others, this shapes a pattern of response to contact through resistance.

Develop. This is a stage variable in how we work with the child, youth, and young adult as a human system to grow, mature, and develop the skills to live, learn, think, and respond to contact. When person is thought of in connection with being human, it changes the way they are built up for contact in the home, school, neighborhood, and workplace network.

Education. This term is related to family, government, and business affairs through a systems theory developed by Dr. Dolores Slaton. When a person has a sense of self, other people (family), and the environment (home, school, neighborhood, and workplace), they have been built up to cooperate as a skill.

Emotion. The subject in self science as feelings within the person and those that erupt in other relationships (Goleman 1995). In human systems research, it is the energy that results from feeling things that influence sense and receive paths. When a person has control over the human system, the brain, body, and senses are set up as a learning system to manage the flow of feelings to and from the brain.

Emotional. This is a human systems domain term. When a person has control over the human system, the brain, body, and senses are set up to manage how it feels to feel: feelings.

Event. "Any occurrence or phenomenon that has a definite beginning and end and that involves or produces change" (American Psychological Association 2007). When a person moves in pursuit of a goal, the act of moving through the event produces an affect. This is a social experience connected with a person, place, or issue, which sets up the act to learn from the affect.

Environment. This is a human, social, and ecological term used to refer to the self, other people, and surroundings as environments. When a person is becoming a human being in the womb, they are an environment within the environment of another person (parent) within the context of other environments.

Environmental. This is a human systems domain term. When a person has to learn how to live, it means they have to learn how live with self, other people, and homes, schools, neighborhoods, and workplaces as environments.

Experience. A core variable in the Progressive Investing practice to learn how to live each day to become more informed. It is the experience, which focuses how the participant learns from the action that is taken to produce feelings that lead to thought and a more reflective capacity. This is the goal of all acts to live, learn, think, and respond. When a person takes action to learn, it is to learn how to move through feelings of contact.

Family. This term is related to education, government, and business affairs through a systems theory developed by Dr. Dolores Slaton. When a person has a sense of self, other people (family), and the environment (home, school, neighborhood, and workplace), they are more competent to lead and compete. The concept of family is related to the family system and family-centered learning.

Family-Centered Learning. A way to set a family up through roles and functions that allow a parent to lead and develop its members, children to functions as learners and participants and to function as a kinship group to live, learn, and act together as a learning team.

Feed Forward. The capacity of the brain to produce responses to contact that can be received or sent forward based on whether the interaction is new or processed feelings.

Feel. A core variable in the act to explore sense data to learn as a human from the experience of contact. To feel a thing, object, or issue means to come in contact with it and to gather a sense of how it feels in relation to the body and brain events.

Feelings. These are produced by bodily contact through the way a thing, object, or issue feels, which forces the flow energy, action, or a reaction.

Feeling System. A human systems term used to teach a child, youth, or young how to recognize the movement of sense data as feelings of contact and interaction between the brain, body and senses. The learner learns how to receive information, process knowledge, respond to experience, and use reflection to build a higher sense of self and the external world.

Free and Appropriate Public School Education. This is a legislative declaration for public school education for children with disabilities at no costs to their parents. When a person has been hurt by major life events that affect how they behave at home, learn in school, and think in their neighborhood their right to participate in free and appropriate public school education may be protected through the Individuals with Disabilities Education Act (CDE 2008).

Goal. This is the effect for which the brain has to be set up to feel. When a person's is set up to live each day to become more informed, they take action to, learn, think, and respond through how it feels to have an aim in mind.

Government. This term is related to family, education, and government affairs through a systems theory developed by Dr. Dolores Slaton. When a person has a sense of self, other people (family), and the environment (home, school, neighborhood, and workplace), they are more cooperative and able to participate and compete.

Green Learning System. A term used by human systems researchers that is virtually synonymous with a human learning systems approach.

Green science and evolving definitions. These refers to a systems approach being developed by Dr. Slaton as a new human system theory and practice that seeks to reduce the gap between functional and dysfunctional states of labor for parents and children in crisis due to poor school experience or from being upset by major life events in man-made home, school, neighborhood, and workplace networks.

Green Technology. A term used in reference to the evolution of the person as a functional system, human system, and hence, reprocessed technology which reflects a use of energy, control, and reprocess cycles to become more aware through brain, body, and sense coordination.

Grow. This is a stage variable in how we work with the child, youth, and young adult as a human system to grow, mature, and develop the skills to live, learn, think, and respond to contact. When person is thought of in connection with being human, it changes the way they are set up to enter and exit contact in the home, school, neighborhood, and workplace network.

Growing up hurt. A functional theme related to major life events in home, school, neighborhood, or workplace networks that affect a person's ability to live, learn, think, and respond to contact and interaction, because of poor cognitive experiences. The process is linked to negative patterns of experience that produce signs of resistance to contact.

Help. A term used alone or in combination with the act to serve in aid to self or another person. In Progressive Investing, the word is used to inspire the helper to help and receive the help of the helpee over time through emerging awareness for its signs of care.

Helpee. The parent, child, or client in need of the service. In Progressive Investing, the word is used to inspire the helpee, to help the helper, help him or her receive the help.

Helper. The service provider, but can be linked to anyone's energy, being used in the situation to improve a participant's response to contact and interaction. In Progressive Investing, the word helper is used to inspire people to want to live each day to learn how to help.

High Need. This is a term that refers to an affect on a person being influenced by emotional stress. When a person is in crisis they have to learn how to experience changes in the way they feel to live through major life events that threaten their sense of self-control.

Home. A symbolic place made by human beings for other human being to live in and to experience. In community-based learning the home is taught as the set up to improve school, neighborhood, and workplace relations. In this sense the home is connected to the school, neighborhood, and workplace as a way to teach participants how to use them to move through them as a network, even though they are disconnected. The home affects how a child may make choices, to live.

Home-Based Learning. This is a learning academy for children, youth, and young adults that need special help to learn how to live through their relationships in the home. When a person has learned to withdraw from parents and siblings it reduces their openness to contact in group events.

Human Learning. This is determined by how the body responds to sense data, and how the brain receives sense data as feelings that may result in thought and reflection.

Human Learning System. This is determined by the way a person is able to receive, organize, and express information as knowledge, experience, and reflection; as components of thought.

Human Science. This is the approach used to learn about the brain, body, and senses to develop a school of thought. A person learns how to move through a problem situation, using science to set up how they live each day to become more informed.

Human System. This is the brain, body, and senses in particular, but how the whole being works in relation to acts to live, learn, think, and respond to contact and interaction as events to function more efficiently. In human systems research, the human system is studied in relation to physics, chemistry, and biology as natural sciences to learn why and how the brain, body, and senses function to produce the capacity and potential to live, learn, think, and respond to tension, stress, and pressure.

Human Systems Thinking. The result of process cycles applied to the flow of feelings as energy through sense, body, and brain activity, which moves contact and interaction that forms as information, knowledge, experience, and reflection to produce structures of thought.

Human Systems Research Domains. This is a systems a general systems theory on how humans learn through relating the things they will feel to human life, which is relative to organic life as having similar processing tasks to live through, and to sustain life (Boulding 1956; Kauffman 1980). When a person learns how to talk to a child's brain, they think about the mechanics of how the body has to be set up to interact as a human ecology that feels things to transfer signs of contact through the senses. We use the words language, cognition, physical, mental, social, environmental, emotional, sensory organization, and green technology; to express these events as process variable in the child's tasks to learn how to live.

Informed. A Progressive Investing term, which means to become more aware.

Information. "A distinction which reduces uncertainty" (Checkland 1999, 315). Our study of information theory refers to the creation of messages and the transmission processes that are involved in the management and control of feelings as signs of tension, stress, or pressure being treated through special process cycles to focus and to respond.

Information Processor. A human systems term, which means that the participant has the capacity and potential to release brain, body, and sense responses at advanced rates, because of the way he or she is set up to practice the act to move feelings through contact and interaction.

Interaction. The transaction, organization, and communication between the effects of contact as information, motives, and drives in the exchange of values tied to personal states and human behavior (Kuhn 1975). It is the connection between contact being sensed, and contact being received, and in the social sense of being observed as mixed energy, action, or feelings. This is the process variable in learning how to learn from contact, used to study the connection between self, other people, and the environment. When a person moves in relation to contact, it is to interact as a physical system.

Language. A communication process used to convey messages through the use of words and talk: written and spoken as a school of thought. This is a human systems domain term. When a person is learning how to live, it means that they are learning how to use noises, sounds, signs, and symbols to read, write, speak, and make sense of numbers.

Learn. This is an action and program variable in the field of experience. When a person is in the act to receive contact, the process of learning from the experience will take place in the home, school, neighborhood, or workplace network.

Learning. This is the act to receive contact in the process to respond and interact with the senses to feel, and focus the flow of feelings through the brain and body. When a person experiences contact, the brain, body, and senses interact with it to learn how to think through the why it feels.

Learning and Support Services. This is the cause, effect, and relationship variables behind the aim to improve the lives of children and families in crisis; and the home to school experiences of poor, minority, high-need and special-need children, youths, young adults, and parents. When a person is growing up hurt by major life events, or has not been set up for classroom learning, they need a human, cognitive, and behavioral plan to learn how to live at home to prepare to learn how to learn in school.

Live. This is the act to feel contact and a program variable in the field of experience. When a person takes action it is to feel for signs of life in relation to self, other people, and environments.

Major Life Events. This is the phrase used to assess a child, youth, young adult, or parent for learning, behavior, or conduct deficits in connection with their home, school, neighborhood, and workplace experiences. When a person has been hurt, the search for the affect can be found in relation to how they are making contact, sustaining interaction, and living cooperatively with the people who share these activities with them.

Mature. This is a stage variable in how we work with the child, youth, and young adult as a human system to grow, mature, and develop the skills to live, learn, think, and respond to contact. When person is thought of in connection with being human, it changes the way they move through contact in their home, school, neighborhood, and workplace network.

Mechanics. This is a technical term used to mean how the brain, body, and senses, need to be set up to work with man-made programs. When a person's brain, body, and senses have been set up to learn in school, their academic work is in response to the programs affect on feelings.

Mental. This is a human systems domain term. When a person is learning how to learn, it means that they are learning how to use the brain to think.

Mental Systems. This is a human systems science term. When a person's brain, body, and senses work in relation to each role to live, learn, think, and respond to contact it is the brains job to lead, which cause the process to fuse as a whole.

Neighborhood. A symbolic place where human experience is plainly linked to child play and development, friendships, kinship groups, free enterprise, and encompasses the home and school. The neighborhood affects the way a child, youth, or young adult learns to think. (Observe how the word neighborhood is connected to the word home above.)

Neighborhood-Based Learning. This is a learning academy for children, youth, and young adults that need special help to learn how to think through their neighborhood relationships. When a person has learned to isolate from friends and neighbors it reduces their will to participate in social events.

Organization. This is a functional dynamic of human systems research to the study of the self, other people, and environment to learn from their action. When a person moves in relation to contact, how and why they feel or fail to feel things become more known through the way their behavior patterns are revealed.

Other People. This is a human systems research term used to refer to people outside the realm of self. When a person interacts with another person, they experience the affect of contact beyond self as a shared event in the social context of environmental learning.

Parent Education and Resource Coordination Services. This is the organization family, medical, educational, mental health, social work, and corrections records and fields to improve learning and support services to children and families in crisis due to learning, behavior, or conduct deficits.

Participation. This is the effect variable in the aim: to move through contact; to interact; and to cooperate; which means to participate in the connection of self to other people, and the environment. When a person takes action to live each day to become more informed, they learn from the experience, how to participate.

Physical. This is a human systems domain term. When a person is learning how to think, it means that they are practicing ways to interact and be seen as social.

Physical System. This is a human systems science term. When a person's brain, body, and senses work in relation to each role to live, learn, think, and respond to contact it is the body's job to function as a whole effect.

Poverty. This is a term used to connect a lack of resources to conditions that may cause learning, behavior, and conduct problems to be more probable and/or expected.

When a person is born into poverty the need to learn how to live with self, other people, and the environment becomes more vital, since the act to learn, think, and respond in public places take shape in competitive structures.

Process. This is a term used in the study of a path of action between a set of elements designed to measure change. When a person takes action to learn how to live in a home, learn in a school, think in a neighborhood, or respond in a workplace they become more flexible and open to the practice of changing with each day.

Process Cycles. This is a term used to connect a group of elements that set up pathways in which to measure changes. When a person can learn how to live through bad situations using: sense, feel, and focus cycles in relation to how they receive, process, and respond to change.

Progressive Investing. The act to live each day to learn how to become more informed.

Progressive Investing and Green Science. The aim to improve sense and receive paths by teaching parents and children how to move through negative feelings with a greater sense of awareness for the consequences in the waste of time, energy, and space when they do not try. Progressive Investing practices focus on energy use, control, and reprocessing. In this sense, Progressive Investing is the act and process to learn from experience how to live each day to become more informed through the use of energy, control, and reprocess cycles.

Progressive Investing Institute of Focused Learning. This is a progressive investing school of thought. When a person learns how to take action to live with the self inside their body, they are learning how to learn with a brain through higher senses of becoming each day.

Progressive Investing Systems. This is the process used to examine the meaning of a set of elements to learn more about how the human system functions. When a person is learning, they are being changed by the affect which produces a more informed response.

Receive. The mental process variable in the act to accept or resist the transfer of sense data.

This term is used in the study of how sense data is moved. When a person comes in contact with an object the brain, body, and senses respond.

Receive Data. The point between the act to accept or resist sense data in the sense, feel, and focus cycle.

Receive, Process, Respond Cycle. The transfer of sense data that has been felt and focused to the mental process and the brain, which may produce a response through thought.

Reducing Family Decline. This phrase is used to set up the goal behind the building human asset movement. When a person learns how to live in a home, they are learning how to be or become a leader and family developer.

Respond. This is the term used to study the path between the act to receive contact, and the practice to process how the contact feels. When a person is set up for academic learning, the brain is open to receiving orders that may cause some signs of discomfort in the preparation of a preferred response to pressure.

School. A symbolic place where human experience is plainly linked to academic learning. School affects how a person learns to participate in social events. (Observe how the word school is connected to the word home above.)

School-Based Learning. This is a learning academy for children, youth, and young adults that need special help to learn how to learn through their relationships. When a person has learned to disconnect from teachers and peers it reduces their will to interact in academic events.

School Failure. This phrase is used in relation to poor school experiences. When a person learns how to learn in a school, they are learning how to interact with people.

School of Thought. A Progressive Investing approach to human development; to become more informed through the study of human, cognitive, and behavior sciences. This is the purpose of teaching people to live, learn, think, and respond as fields of learning. When a person learns how to take action, they are learning to focus the process of thought.

Self. This term refers to the individual, person, or human being. When a person develops a higher sense of self, they are learning about the individual or person inside their body and what it means to be a human being with a brain.

Send forward. Moving energy and action as feelings to and from the brain as messaging between send and receive points based on the basis of the feed.

Sense. The process variable in the social act to accept or resist feelings of contact.

Sense Data. Energy, action, or feelings of contact between the body and another environment.

Sense, Feel, and Focus Cycle. A human systems research term used to describe the process flow energy, action, and feelings undergo to transfer data to the brain.

Senses. This term refers to organs and nerve networks with direct links to the brain. When a person can feel something, it means the senses have made contact with an inner or outer source.

Sensory. This is a human systems domain term. When a person is learning how it feels to be human, it means that they are thinking of ways to cooperate with these feelings moving through their body.

Sensory Organization. The coordination of brain, body, and sense events to move through contact more efficiently. Hence, we ask the parent and child to take action to become more informed from the experience of contact; the parent and child works through the use of their senses to become more aware in regards to the things they see, hear, feel, tastes, and experience as pressure.

Social. This is a human systems domain term. When a person is learning how to respond, it means that they are living to learn how to behave in public places.

Social Issues. This term relates to a pattern of behavior in public places. When a person is learning how to act, it means they are trying to change a pattern of responding.

Special Need. This term is used in reference to people that have been assessed as or with a physical or mental disability. When a person proves to have a learning, behavior, or conduct disorder, they become eligible to receive special services through federal and state laws.

System. "An organized whole made up of components that interact in a way distinct from their interaction with other entities and which endures over some period of time" (Anderson and Carter, 1990, 266).

Systems Feelings. The conceptualization of how the brain, body, and senses make up the core elements in the human act to feel things to live, learn, think, and respond from the experience of interaction. Hence, it is the fusion of action science and involves the conceptualization of the community-based learning context as a field of experience in relation to contact, and with an emphasis on the state of being human and the theory of how the human system functions.

Systems Thinking. This is a way of observing the human system to learn how the brain works to live, learn, think, and respond. When a person is learning how to live, they will experience contact, growth, structure, and control (Checkland 1999).

Taking Action. This term represents the theory and practice of progressive investing, to learn how to live. When a person is learning how to move in relation to contact, they learn from the experience of interaction, to think.

Think. This is an action and program variable in the field of experience. When a person is in the act to receive contact, the process to interact takes hold and the brain shapes all thoughts in response to these feelings.

Thought. A product of the brain—human system body and sense networks.

Unemployment. This phrase is used in relation to a lack of skill sets due to stages of withdrawal from the experience of contact in the home, disconnect from the experience of interaction in school, and isolation from the experience of cooperation in the neighborhood, which affect how well a person prepares to participate in the workplace.

Workplace. A symbolic representation of human experience in relation to school behavior and performance plainly linked to labor and skills in society. When a person learns how to try to perform in home, school, and neighborhood situations, they are better able to respond in the workplace.

Workplace-Based Learning. This is a learning academy for children, youth, and young adults that need special help to learn how to respond through their workplace relationships. When a person has learned to resist contact and interaction with self, other people, and the environment, it reduces their will to focus on workplace events.

References

Ackerman, P.L. (November, 2003). Cognitive ability of non-identity trait determinants of expertise. Educational Research, Vol. 32, No. 8 (pp. 15-20). American Educational Research Association.

Agazarian, Y.M. (1997). Systems-centered therapy for groups. New York: The Guilford Press.

American Psychiatric Association (2002). Diagnostic and statistical manual of mental disorders: DSM-IV-TR. Fourth edition.

Anderson, R.E. & Carter, I. (1990). Human behavior in the social environment: A social systems approach. (4th ed.). New York: Aldine De Gruyter.

Appelbaum, P. (2002). Multicultural and diversity education: a reference handbook. Santa Barbarah: ABC-CLIO.

Argyris, C., Putnam, R., & Smith, D.M. (1985). Action science. San Francisco: Josey-Bass Inc.

Argyris, C. & Schon, D.A. (1974). Theory in practice: increasing professional effectiveness. San Francisco: Josey-Bass Publishers.

Armstrong, T. (2006) The best schools: How human development research should inform educational practice. Alexandria: Association for Supervision and Curriculum Development.

Audi, R. (1999). The Cambridge dictionary of philosophy: Second edition. Cambridge: University Press.

Banks, A.B. (2002). An introduction to multicultural education. Boston: Allyn and Bacon.

Banks, J.A., & Banks, C.A.M. (2003). Multicultural education: issues and perspectives. (4th edition). New York: John Wiley & Sons.

Barten, A.C., Drake, C., Perez, J.G., St. Louis, K., & George, M. (2004). Ecologies of parental engagement in urban education. Educational Researcher, Volume 33, No. 4, p. 3-12 May, 2004.

Bentz, V.M., & Shapiro, J.J. (1998). Minful inquiry in social research. Thousand Oaks: Sage Publications.

Blankenhorn, D. (1995). Fatherless America: Confronting our most urgent social problem. New York: Harper Perennial, a Division of Haper Collins Publishers.

Bloom, B.S., (ed). (1956). Taxonomy of educational objectives: the classification of educational goals. London: Longmans, Green and Co. LTD.

Boulding, K. (April, 1956). General system theory: the skeleton of science. Management Science. Volume 2(3).

Boynton, M., & Boynton, C. (2005). The educator's guide to preventing and solving discipline problems. Alexandria: Association for Supervision and Curriculum Development.

California Codes (2006). Welfare and Institutions Code, Section 5600-5623.5 (Available on line: *www.leginfo.ca.gov/ca.bov/cgi-bin/displaycode?section=wic&group=05001*-06000&file=56.

California Corrections Standard Authority (2005). Juvenile detention profile survey (4th Quarter). Corrections Standards Authority, Facilities Standards and Operations Division: Sacramento, California.

California Department of Education (2006). State schools chief Jack O'Connell Grants Millions for teacher recruitment and student learning. (Available online: *www.cde.ca.gov*).

California Department of Education (2006). Statewide summary of expulsions and recommended expulsions by education codes sections. Educational Options Office (Available on line: *www.data1.cde.ca.gov/dataquest/expulsion/EdCodeSt.asp?cYear=2-3-4&cChoice=EdCodeSt&pageno=1*).

California Prisoners and Parolees (2004). California Department of Correction and Rehabilitation: Sacramento California.

California Department of Education (2005). High performing schools initiative: A white paper on improving student achievement in California high schools. California Department of Education, February 2005. (Available online: *www.cde.ca.gov/eo/in/se/yr05highschoolwp.asp*).

California Department of Education (2006). Statewide dropout reports. Educational Demographics Unit, March 27, 2006. (Available online:

Capra, F. (2002). The hidden connections: A science for sustainable living. New York Anchor Books.

Capra, F. (1999). The tao of physics: An exploration of the parallels between modern physics and eastern mysticism. Boston: Shambhala Publications.
Checkland, P. (1999). Systems thinking, systems practice: Includes a 30-year retrospective. Chester: John Wiles & Sons, LTD.
Chidsey, R.B., & Steege, M.W. (2005). Response to Intervention: Principles and strategies for effective practice. New York: The Guilford Press.
Children's Defense Fund. (2007). America's cradle to prison pipeline. A Children's Defense Fund Report. October, 2007.
Chomsky, N. (2002). On nature and language. Cambridge University Press.
Crystal, D. (2003). A dictionary of linguistics and phonetics. (Fifth edition). Malden: Blackwell Publishing.
Cuban, L. (2004). The blackboard and the bottom line. Why schools can't be businesses. Cambridge: Harvard University Press.
Dattilio, F.M., & Freeman, A. (2000). Cognitive-behavioral strategies in crisis management. New York: Guilford Press.
Denzin, N.K. (1989). Interpretive interactionism. Newbury Park, CA: Sage.Dictionary of Science (2003). A dictionary of science. New York: Oxford University Press.
Dewey, J. (1910/1991), How we think. New York: Prometheus Books.
Dewey, J. (1922). Human nature and conduct: An introduction t social psychology. New York: Macmillan Company.
Dewey, J. (1927). The public & its problems. New York: Henry Holt and Company.
Dewey, J. & Bentley, A.F., (1949). Known & the unknown. Boston: Beacon Press.
Dunn, C., Chambers, D., & Rabren, K. (2004). Variables affecting student's decisions to drop out of school. Remedial and Special Education, Volume 25, Number 5, September/October p. 314-323.
Educational Leadership (2004). Awareness of high-needs students: the best of Educational Leadership 2003-2004.
Einstein, A. (1961). Relativity: The special and general theory. New York: Three River Press.
Eisenhart, M., & Towne, L. (October, 2003) Contestation and change in national policy on "scientifically based" education research. (p. 31-38). In Education Researcher, Volume 32. Number 7. American Educational Research Association.

Erickson, M.F., & Riemer, K.K. (1999). Infants, toddlers and families: A framework for support and intervention. New York: The Guilford Press.

Erikson, E.H. (1963). Childhood and society. New York: W.W. Norton & Company.

Gabriel, J.G. (2005). How to thrive as a teacher leader. Alexandria: Association for Supervision and Curriculum Development.

Gaines, B.R. (2004). (2004). Organizational knowledge acquisition. In Holsapple, C. (Ed.). Handbook on knowledge management 1: Knowledge Matters. Springer: New York.

Garcie, D.C., & Hasson, D.J. (2004, Spring/Summer). Implementing family literacy programs for linguistically and culturally diverse population: key elements to consider. The School Community Journal, Vol. 14, No. 1., (pp 113-137).

Gina Biancarosa (October, 1995, p. 16)

Goleman, D. (1995). Emotional intelligence: Why it can matter more than IQ. New York Bantam Books

Hartman, T., (2006). Screwed: the undeclared war on the middle class, and what we can do About it. San Francisco: Barrett-Koehler Publishing, Inc.

Hopson, D.P., & Hopson, D.S. (1990). Different and wonderful: Raising black children in a race-conscious society. New York: Simon & Schuster.

Hussar, W.J. and Bailey, T.M. (2009), Projections of Education Statistics to 2018 (NCES 2009-062). National Center for Education Statistics. Institute of Education Sciences, U.S. Department of Education, Washington, D.C.

Kauffman, D. (Jr.) (1980). Systems one: an introduction to systems thinking. St Paul: Future Systems, Inc.

Kuhn, A. (1975). Unified social science: A system-based introduction. Homewood: The Dorsey Press.

Kuhn, T. (1970). The structure of scientific revolutions. Chicago: University of Chicago Press.

Kendall, D. (2001). Social problems. (2nd ed.). Boston: Allyn and Bacon.

Labaree, D.F. (1997). How to succeed in school without really trying: the credential race in American Education. Yale University.

Leonardo, Z. (August/September, 2004). Critical social theory and transformative knowledge: the function of curriculum in quality education. Educational Research. American Educational Research Association, Vol.33, No. 6, (pp. 11-18).

Letrello, T.M., & Miles, D.D. (2003). The transition from middle school to high school: students with and without learning disabilities share their perceptions. Volume 76, No. 4, March/April 2003. The Clearing House.

Lippman, L., Atienza, A., Rivers, A., & Keith, J. (September, 2008). A developmental perspective on college and workplace readiness. Child Trend. Bill and Melinda Gates

Logan, S.M.L., Freeman, E.M., & McRoy, R.G. (1990). Social work practice with black families: a cultural specific perspective. New York: Longman.

Maslow, A.H. (1966). The psychology of science: a reconnaissance. New York: Van Norstrand & Reinhold Company.

Maslow, A.H. (1968). Toward a psychology of being. (2nd ed.) New York: Harper & Row Publishing.

Maslow, A.H. (1970). Motivation and personality. New York: Harper & Row Publishing.

Maslow, A.H. (1998). Maslow on management: With Deborah C. Stephens and Gary Heil. New York: John Wiley & Sons, Inc.

Marzano, R.I., Waters, T., & McNulty, B.A. (2005). School Leadership that works: From research to results. Alexandria: Association for Supervision and Curriculum Development.

McKeen J.D., & Staples, S.D. (2004). Knowledge managers: who they are and what they do. In Holsapple, C. (Ed.). Handbook on knowledge management 1: knowledge matters. Springer: New York.

Meier, D., Kohn, A., Hammond, L.D., Sizer, T.R., & Wood, G. (2004). How the no child behind act is damaging our children and our schools: many children left behind. Boston: Beacon Press.

Maxwell, J.A. (2004). Causal explanation, qualitative research, and scientific inquiry in education. Educational Researcher, Vol. 33, No. 2, p. 31-11.

Nieto, S. (1996). Affirming diversity: the sociopolitical context of multicultural education. New York: Longman Publishers.

O'Shea, M.R. (2005). From standards to success: A guide for school leaders. Alexandria: Association for Supervision and Curriculum Development.

Paulo, F. (2004). Pedagogy of the oppressed: 30th anniversary edition, translated by Myra Bergman Ramos, with an introduction by Donaldo Macedo. New York: Continuum.

Peterson, J.L. (2000). From rage to responsibility: Black conservative Jesse Lee Peterson and America today. St. Paul: Paragon House.

Piaget, J. (1952). The origins of intelligence in children. New York: International Universities Press.

Pierangelo, R. (Ed.), (2003). The special educator's book of lists. (2nd edition). San Francisco: Josey - Bass.

Popham, J.W. (March, 2005). Instructional quality: collecting credible evidence. Educational Leadership.

Prestwich, D.L. (2004, Spring/Summer) Character education in American schools. The School Community Journal. Vol. 14, No. 1, pp 139-154).

Progressive Investment Report (2002). Black child development: research on how Black youths need to learn to live in society. Save Our Youth, the Next Generations, Sacramento: Progressive Investment Group Press.

Progressive Investment Report (2004). SOY Family Leadership Academy: Building human assets process for attacking urban decay; Research-based practices. SOY, the Next Generations, Sacramento: Progressive Investment Group Press.

Progressive Investment Report (2005). SOY Family Leadership Academy: Research-based practices; in partnership with the Progressive Investing Institute of Focused Learning.

Save Our Youth, the Next Generations, Sacramento: Progressive Investment Group Press.

Progressive Investment Report (2006). Evidence-based reports to our public schools. Why poor, minority, high-need and special-need children cannot learn. SOY, the Next Generations, Sacramento: Progressive Investment Group Press.

Progressive Investment Report (2007-2008). Save Our Youth the Next Generations: Help to improve our public schools. Sacramento: Progressive Investment Group Press.

Public Advocates (2006). Making rights real. Presented by Liz Guillen, Director of Legislative & Association for Supervision and Curriculum Development: Vol. 62, No. 6.

Ray, O., Ksir, C. (1990). Drugs, society, and human behavior. (6th ed.) St. Louis: Mosby.

Reason, P., (Ed.). (1994). Participation in human inquiry. London: Sage Publications.

Reeves, D.B. (2006). The learning leader: How to focus school improvement for better results. Alexandria: Association for Supervision of Curriculum Development.

Reyes, G.I., (Ed.), (2001). A portrait of race and ethnicity in California: An assessment of social and economic well-being. San Francisco: Public Policy Institute of California.

Rowley, L.L. (2004) Dissecting the anatomy of African-American inequality: the impact of racial stigma and social origins on group status and college achievement. Educational Researcher, Volume 33, No. 4, May 2004.

Ruggiero, V.R. (1998). Beyond feelings: A guide to critical thinking. Mountain View: Mayfield Publishing Company.

Sagor, R. (1992). How to conduct collaborative action research. Alexandria: Association for Supervision & Curriculum Development.

Scherer, P.L., Pennell, G.S., Lyons, C., & Fountes, I. (October, 2005). Becoming an engaged reader. Educational Leadership (p. 24-29). Foundation.

Segal, J., & Yahraes, H. (1978). A child's journey: forces that shape the lives of our young. New York: McGraw—Hill Book Company.

Sizer, T.R. (2004). The red pencil: Convictions from experience in education. New Haven: Yale University Press.

Slaton, C.K. (2002). Progressive investing in systems feeling: A human approach on how to help black families succeed at home, at school, at work, and in their neighborhoods. Improving family relations. Sacramento: Progressive Investment Group Press.

Slaton, C.K. (2004). Progressive investing in mind and body: Progressive Investment Report Series, Progressive Investing Institute of Focused Learning, a systems feeling perspective. Sacramento: Progressive Investment Press.

Slaton, C.K. (2009). Education and science: how the body lives, how the brain learns, how the human system thinks, how human systems research responds. A Progressive Investing Perspective. Sacramento: Xlibris Corporation.

Slavin, R.E. (2004). Education research can and must address "what works" Questions. Educational Researcher, Vol. 33, No. 1, p. 27-28.

Sousa, D.A. (2001). How the brain learns: a classroom teachers guide. (2nd edition). Thousand Oaks: Corwin Press.

Stringer, E.T. (1999). Action research (Second Edition). Thousand Oaks, CA: Sage Publications.

Sullo, B. (2007). Activating the desire to learn. Alexandria: Association for Supervision and Curriculum Development.

Swanson, C.B. (2005). Who graduates in California: Minority students lag behind, effects of segregation persist. The Urban Institute Education Policy Center, March 2005.

Szsaz, T. (1990). Insanity: the idea and its consequences. New York: John Wiley & Sons.

Taba. H. (1932). The dynamics of education. A methodology of progressive educational thought.

Kegan Paul, Trench, Trubner & Co., Ltd. Reprinted in 1999 by Routledge: London.

Tatum, A.W. (2006). Engaging African American males in reading: by providing meaningful reading material and encouraging honest debate, teachers can help African American adolescent males embrace the power of text. Educational Leadership, Volume 63, No. 5 February 2006.

Taylor, M. (January, 2009) Improving academic success for economically disadvantaged Sacramento: Legislative Analyst Office.

U.S. Department of Education (2004). Special education-enrollment by grade and disability. Special Education Division (Available on line: *www.ed-daata.k12.ca.us/welcome.asp*).

U.S. Department of Education National Center for Education and Statistics (2009). Crime, violence, discipline, and safety in U.S. Public Schools: Findings from the School Survey on Crime and Safety: 2007-08.

U.S. Department of Labor (2004). Bureau of Labor Statistics (Available on line: *www.bls.bov/cps/minwage2004_htm*).

U.S. Government Accountability Office (2007). Report to the Chairman, Committee on Ways and Means, House of Representatives. African American children in foster care. Additional HHS assistance needed to help states reduce the proportion in care. GAO-07-816.

Watts, I.E., & Erevelles, N. (2004). These deadly times: reconceptualizing school violence by using critical race theory and disability studies. American Educational Research Journal, Volume 41, No. 2. p. 271-299, Summer 2004).

Wells, G. (1999). Dialogic inquiry: towards a sociocultural practice and theory of education. Cambridge: Cambridge University.

West, T.G. (1997). In the mind's eye: Visual thinkers, gifted people with dyslexia and other learning difficulties, computer images and the ironies of creativity. New York: Prometheus Books.

Wheatley, M.J. (1999). Leadership and the new science. Discovering order in a chaotic world. Berrett-Koehler Publishers: San Francisco.

Wilson, A. N. (1992). Awakening the natural genius of black children. New York: Afrikan World InfoSystems.

Wortham, S. (Fall, 2004). The interdependence of social identification and learning. American Educational Research Journal. Vol. 41, No. 3 (pp.715-750).

Vygotsky, L.S. (1978). Mind in society: The development of higher psychological processes. Cambridge: Harvard University Press.

Index

A

action, 207, 216, 219, 229, 242, 370-71
affect, 107, 230, 240, 248, 255-56, 270
aim, 230, 300, 313, 325-26
anger, 103-4, 106-7, 123-25, 143-46, 321-22, 348-50

B

behavior, 20-22, 188-91, 215-17, 273-75, 282-84, 319-21
body, 70, 84, 227-28, 230, 254, 256
brain, 15-22, 75-80, 82-91, 109-15, 285-99, 309-21
building human asset meeting, 51, 55, 67, 193-94, 196
Building Human Asset Process, 55, 108, 123, 192
Building Human Assets, 4, 55, 211, 376
business, 47, 108, 160, 190, 209, 242-43

C

capacity, 78-80, 94-97, 114-15, 144-46, 155, 176-78
child developer, 13, 149-50, 184, 345-46

cognition, 77-78, 80
community-based learning, 172
Community-Based Learning Certificate of Achievement, 194, 197-98, 351
contact, 18-24, 69-80, 99-111, 247-56, 279-86, 317-22
cooperation, 92-93, 139, 168, 210, 280-83, 322
crisis, 27-28, 152-54, 194-98, 200-203, 209-11, 213-17

D

data, 44, 251, 257, 263, 368
delinquency, 14, 17, 183-84, 196-97, 210, 252
disappointment, 103, 108, 112, 129, 145, 323-24
dropouts, 39, 41, 43-44, 53

E

education, 160, 174, 197, 221
emotion, 102, 104, 106, 109, 113, 140
energy, 18, 79, 270-71, 309-12, 322, 337-44
environment, 96, 223, 229, 240

event, 107, 219, 230, 239, 248, 253-54
experience, 71-80, 106-14, 173-79, 181-88, 285-89, 309-17

F

family, 27-31, 117-20, 122-45, 147-50, 152-53, 191-98
  cognitive state, 127
  education and, 152
  emotional state, 143
  environmental state, 140
  language state, 124
  leadership, 117, 148, 164
  literacy, 30, 33
  mental state, 134
  physical state, 130-31
  school failure and, 39
  social state, 137
family-centered learning, 47, 135-37, 139, 141, 153, 248
fear, 24, 30-31, 103-4, 106-8, 144-45, 320-22
feeling, 4, 99, 150-51, 173, 182, 230
feeling system, 4
focus, 228, 301, 310, 319, 325, 328

G

goal, 107, 230, 240, 242, 248, 253-54
government, 38, 46, 65, 168-69, 174, 242-43
green technology, 112-13, 149, 153, 192
growing up hurt, 217, 344

H

hate, 50, 129
help, 55, 253, 260, 295, 376
helper, 229, 253
home, 46-51, 92-98, 185-98, 250-63, 276-83, 344-48
human learning, 4, 18, 230, 238, 243, 247-48
human system, 22-23, 70-73, 167-69, 202-6, 267-68, 270-74
Human Systems Research (HSR), 46-47, 90, 94, 115, 180, 204
human systems science, 4, 13, 57, 108, 191, 338
hurt, 22-25, 118-23, 147-49, 193, 296, 344-51

I

identity, 45, 101, 108, 157-58, 162-63, 169
interaction, 19-24, 82-85, 91-94, 118-19, 125-28, 280-83

K

knowledge, 46-47, 150, 162-63, 174-75, 297, 319-20

L

language, 70-71, 73-74, 78, 80
leadership, 55, 117-18, 148, 170-71, 191-93, 375-79
learner, 16-18, 117-18, 154, 163-64, 309-11, 313-19
learning, 4, 197-99, 210, 247-48, 325, 339

learning disabilities, 21, 45

M

mental, 86
mind, 46-47, 136-39, 267-68, 270-74, 285-87, 295-99

N

No Child Left Behind Act, 40, 49, 68, 221

P

parent education, 17, 33, 191, 197, 200-201, 203
parenting, 17, 24-25, 30, 120, 150, 175-79
parent-school partnership and accountability, 67
participant, 172, 179
participation, 49, 54, 58, 91-93, 208-9, 230
physical, 81, 83, 91
pressure, 101-2, 155-56, 191-92, 249-51, 260, 273-74
Progressive Investing, 15-16, 75-81, 93-97, 188-92, 288-92, 312-18

R

reflection, 217, 332, 339
resistance, 23-24, 55-58, 105-6, 178, 296, 319-20
respond, 243, 250, 339, 350

S

safe places, 181
science, 108, 110, 160, 221, 264, 299
sense, 148, 150, 228, 327
sensory organization, 108, 110, 115, 147-48, 216
social, 91, 104, 115, 143, 161
space, 18, 81, 173, 184, 273
special acts, 179
special education, 181
special places, 181
special practice, 179
special science, 180
special theory, 179
strategy, 203
stress, 101, 273, 289, 345
structure, 199
system, 202, 232

T

talk, 261-62, 317
think, 330
thought, 134, 175, 250
time, 18, 265, 273
training, 193, 263

W

words, 76, 83, 111, 126, 264, 305
workplace, 46-48, 94-97, 163-64, 200, 261-63, 276-77

www.ingramcontent.com/pod-product-compliance
Lightning Source LLC
Chambersburg PA
CBHW031817170526
45157CB00001B/86